Success guides

Leckie×Leckie

D1146832

Intermediate 1
Chemistry

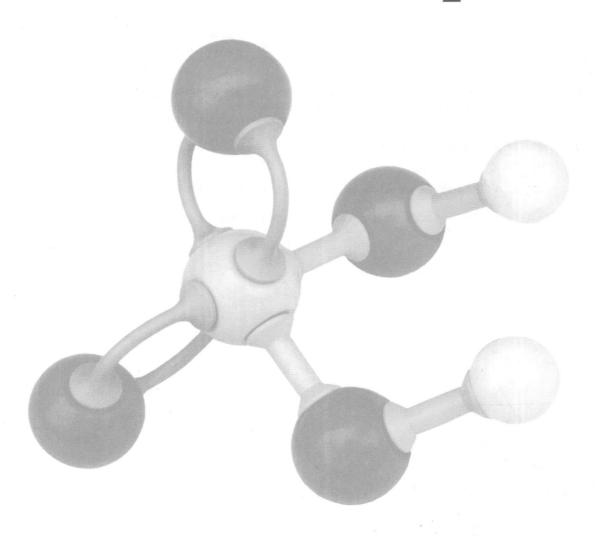

Archie Gibb ✕ David Hawley

ISBN 978-1-84372-610-4

Published by
Leckie & Leckie Ltd, 3rd floor, 4 Queen Street, Edinburgh, EH2 1JE
Tel: 0131 220 6831 Fax: 0131 225 9987
enquiries@leckieandleckie.co.uk   www.leckieandleckie.co.uk

Special thanks to
Helen Bleck (copy-editing), Graham Cameron Illustration/Graeme Wilson (illustration),
Ken Vail Graphic Design (layout and illustration), Tara Watson (proofreading)

Acknowledgments
Leckie & Leckie would like to thank the following for their kind permission to reproduce their material:
Topham Picture Point (photo, p23); Solway Recycling (photo, p53);
Food and Agriculture Organisation of the United Nations (photo, p64)

Leckie & Leckie makes every effort to ensure that all paper used in our books is made from wood pulp obtained from well-managed forests.

A CIP Catalogue record for this book is available from the British Library.

Leckie & Leckie Ltd is a division of Huveaux plc.

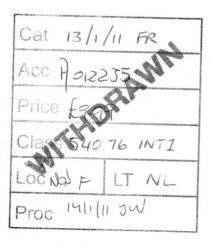

Contents

Intermediate 1 Chemistry

Course Structure

The Intermediate 1 Chemistry course is divided into three units:

Unit 1: Chemistry in Action   Unit 2: Everyday Chemistry   Unit 3: Chemistry and Life

Assessment (Exams and tests)

There are two types of assessment – **internal** and **external**.

For each of the three units, the **internal assessment** consists of a NAB test.

You have to gain at least 18 marks out of a possible 30 marks to pass. In the unlikely event of you failing a NAB, you will be given the opportunity to sit another **one** on the same unit.

Practical Abilities are also assessed internally. This consists of writing a report on **one** of the PPAs from Unit 1.

The **external assessment** consists of an examination which lasts for 1 hour and 30 minutes.

The exam paper is divided into two sections:
- **Section A** which is worth **20 marks** is made up of 20 multiple-choice questions
- **Section B** which is worth **40 marks** is the written part of the paper. In this section approximately 4 marks are allocated to questions based on any of the nine PPAs.

The course award is graded A, B, C or D depending on how well you do in the external examination. In order to gain the course award, you must also pass the three NABs (one for each unit) and complete a PPA report up to the standard required by the SQA.

The Structure of this Success Guide

This Success Guide has been written specially for the Intermediate 1 Chemistry course and our aim is to help you achieve success in the final exam by providing you with an enjoyable learning experience.

The book is divided into the three units of the course and within each unit there is a two-page spread on each of the sub-sections.

Each two-page spread
- covers the content of the sub-section in a manner that will not only help you learn the key ideas but will allow you to get a good understanding of them as well. There are also lots of coloured pictures and diagrams to aid your understanding.
- contains **'Top Tips'**. These flag up important pieces of knowledge that you need to remember and important things that you must be able to do.
- has a **Quick Test** containing questions designed to test your knowledge and understanding of the content. Answers are also given so that you can monitor your progress.

At the end of each unit there is a section devoted to the three PPAs of that unit. It covers in detail the 'Aim', 'Procedure', 'Results' and 'Conclusion' for each PPA. It also includes a section labelled 'Points to note' in which important aspects of the PPAs are highlighted. In addition there are 'Quick Tests' with questions relating directly to the PPAs.

Towards the end of the book, there is a glossary. This is like a chemical dictionary and it contains definitions of key words that you need to know.

There is a final section on 'Chemical tests' where all the common tests you have to know are drawn together and summarised.

Using the Success Guide

It is recommended that you use this Success Guide throughout the year. After each chemistry lesson you should read the appropriate section in the book and make sure you learn and understand the key facts. You should also do the 'Quick Test' and this will allow you to see the progress you are making.

The Guide will also prove extremely useful in the lead up to the exam when your revision will be in full swing.

Elements

Elements and the Periodic Table

Everything in the world is made from about 100 elements, each of which has a name and a chemical symbol.

Chemists have arranged all the elements in the Periodic Table. The names and symbols of the different elements are given in the Periodic Table on page 8 of your Data Booklet. Each element has also been given a number called the atomic number. The elements are arranged by increasing atomic number. For example, the first element in the Periodic Table is hydrogen. Hydrogen has the symbol H, and its atomic number = 1.

The second element is helium. Helium's symbol is He and its atomic number = 2.

Hydrogen and helium are in the same **horizontal row** or **period** in the Periodic Table.

The element sodium sits below lithium in the Periodic Table. Sodium and lithium are said to be in the same **vertical column** or **group** in the Periodic Table.

It is important to write the symbols of the elements correctly. The first letter must be a capital letter and the second letter (if the symbol has two letters) must be lower case. For example, the element cobalt has the symbol Co. If this is written wrongly as CO then it means carbon (C) joined to oxygen (O) which is known as carbon monoxide, which is very different from cobalt.

Top Tip
Remember that the Periodic Tables in your Data Booklet have the names and symbols of the elements.

Solids, liquids or gases?

Most chemical elements are solid at room temperature.

Two elements, mercury (Hg) and bromine (Br) are liquid at room temperature.

Some elements are gases at room temperature, for example hydrogen, oxygen, nitrogen, fluorine and chlorine. All the elements in Column 0 of the Periodic Table are gases.

Top Tip
Remember that mercury and bromine are the only elements which are liquid at room temperature.

Metals or non-metals?

Look at the Periodic Table on the opposite page. Elements shown below the red line are metals. You can see that there are many more metal elements than non-metal elements. This is similar to the Periodic Table on page 8 of the Data Booklet.

Naturally occurring or made by scientists?

Many elements such as gold, silver and copper have been known for a long time and are found naturally either as the element or in compounds.

Look at your Data Booklet to find out which elements have been discovered most recently. Elements after uranium (number 92) have been made by scientists and do not occur naturally. This can be seen on the Periodic Table below.

Key

Atomic number
Symbol

- naturally occurring
- gases
- liquid
- made by scientists

non-metals
metals
transition metals

Column 1	Column 2											Column 3	Column 4	Column 5	Column 6	Column 7	Column 0
1 H																	2 He
3 Li	4 Be											5 B	6 C	7 N	8 O	9 F	10 Ne
11 Na	12 Mg											13 Al	14 Si	15 P	16 S	17 Cl	18 Ar
19 K	20 Ca	21 Sc	22 Ti	23 V	24 Cr	25 Mn	26 Fe	27 Co	28 Ni	29 Cu	30 Zn	31 Ga	32 Ge	33 As	34 Se	35 Br	36 Kr
37 Rb	38 Sr	39 Y	40 Zr	41 Nb	42 Mo	43 Tc	44 Ru	45 Rh	46 Pd	47 Ag	48 Cd	49 In	50 Sn	51 Sb	52 Te	53 I	54 Xe
55 Cs	56 Ba	57 La	58–71 / 72 Hf	73 Ta	74 W	75 Re	76 Os	77 Ir	78 Pt	79 Au	80 Hg	81 Tl	82 Pb	83 Bi	84 Po	85 At	86 Rn
87 Fr	88 Ra	89 Ac	90–103														

* at 28 atmospheres
† sublimes

57 La	58 Ce	59 Pr	60 Nd	61 Pm	62 Sm	63 Eu	64 Gd	65 Tb	66 Dy	67 Ho	68 Er	69 Tm	70 Yb	71 Lu
89 Ac	90 Th	91 Pa	92 U	93 Np	94 Pu	95 Am	96 Cm	97 Bk	98 Cf	99 Es	100 Fm	101 Md	102 No	103 Lr

Properties and uses of the elements

All the elements in a column of the Periodic Table have similar chemical properties. This means that they take part in the same types of chemical reactions and react in similar ways. For example, lithium, sodium and potassium are all in Column 1 and all react vigorously with water. The elements in Column 0, helium, neon, argon, etc, are all very unreactive gases.

The uses of the elements are linked to their properties. Your Data Booklet page 5 gives information about the uses and properties of some metals.

If you do not have your own copy of the Intermediate1 Data Booklet, you can download it from www.sqa.org.uk.

Quick Test

1. What is the approximate number of chemical elements?

2. Name and write the symbols for elements with atomic numbers
   a) 14   b) 17   c) 20   d) 29

3. In which column of the Periodic Table are all the elements gases?

4. Write down the symbols for
   a) helium   b) gold   c) copper
   d) sodium

5. Name two elements that are not naturally occurring and have been made by scientists.

6. Write down a use for
   a) copper   b) aluminium.

Compounds and mixtures

Compounds

Every element is made up of very small particles called atoms. Atoms of different elements are different.

Compounds are formed when elements react together. A **compound** is a substance which contains two or more elements joined together. Water is the most common compound. It contains the elements hydrogen and oxygen joined together.

You probably know that water has the chemical formula H_2O. This means two hydrogen atoms joined to one oxygen atom.

You may have seen magnesium burning in air. When this happens the magnesium is reacting with the oxygen in the air. The white powder which remains at the end is the compound magnesium oxide.

We say that the elements magnesium and oxygen have joined together to make the compound magnesium oxide.

Magnesium burning in air with a very bright white light

You can see this as pictures of atoms below.

Magnesium is an element. Note how all magnesium atoms are the same.

Oxygen is an element. Note how all the oxygen atoms are the same as each other but are different from the atoms of magnesium.

Magnesium oxide is a compound. Magnesium atoms have joined to the oxygen atoms.

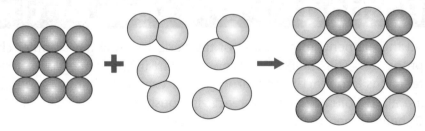

magnesium
(an element)

oxygen
(an element)

magnesium oxide
(a compound)

Naming compounds

Compounds with names ending in **-ide** contain the two elements indicated in the name.

Magnesium ox**ide** contains the two elements magnesium and oxygen.

Sodium chlor**ide** contains the elements sodium and chlorine.

Compounds with names ending in **-ate** or **-ite** also contain the element **oxygen**.

Sodium carbon**ate** contains the elements sodium, carbon **and oxygen**.

Copper sulph**ate** contains the elements copper, sulphur **and oxygen**.

Sodium chlor**ate** contains sodium, chlorine **and oxygen**.

Sodium sulph**ite** contains sodium, sulphur **and oxygen**.

Top Tip
Remember: if the name of the compound ends in **-ite** or **-ate**, then the compound also contains oxygen.

Mixtures

When two substances are **mixed together** but **don't actually react** with each other we say that a **mixture** has been formed. **Common mixtures** include **air** and **crude oil**.

Crude oil is a mixture of different hydrocarbons.

Air is a mixture of gases. Air is approximately 80% nitrogen and 20% oxygen.

The test for oxygen is that it relights a glowing splint.

Although pure oxygen relights a glowing splint, air does not. This is because there is not enough oxygen in the air for air to relight a glowing splint.

Top Tip

Air is a mixture containing approximately 80% nitrogen gas and 20% oxygen gas.

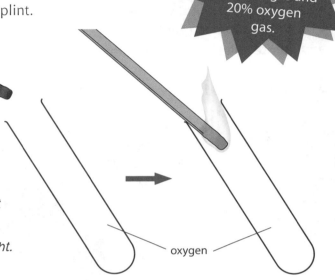

Pure oxygen relights the glowing splint but since air is only 20% oxygen the glowing splint would not relight.

oxygen

Quick Test

1. What is a compound?
2. What is the difference between a compound and a mixture?
3. Copy and complete the table opposite showing the elements present in the compounds.
4. What is the test for oxygen gas?
5. Why does air not relight a glowing splint?
6. What is the main gas in the air?

Name of compound	Names of elements present
Magnesium chloride	
Sodium oxide	
Copper carbonate	
Magnesium nitrate	
Sodium chlorate	
Calcium phosphate	
Copper oxide	
Iron sulphide	
Silver chloride	
Sodium carbonate	

Answers 1. A compound is a substance which contains two or more elements joined together. **2.** In a mixture the substances are not joined together. **3.**

Name of compound	Names of elements present	Name of compound	Names of elements present
Magnesium chloride	Magnesium and chlorine	Calcium phosphate	Calcium, phosphorus and oxygen
Sodium oxide	Sodium and oxygen	Copper oxide	Copper and oxygen
Copper carbonate	Copper, carbon and oxygen	Iron sulphide	Iron and sulphur
Magnesium nitrate	Magnesium, nitrogen and oxygen	Silver chloride	Silver and chlorine
Sodium chlorate	Sodium, chlorine and oxygen	Sodium carbonate	Sodium, carbon and oxygen

4. Oxygen relights a glowing splint. **5.** There is not enough oxygen in air to relight a glowing splint. **6.** Nitrogen is the main gas in the air (about 80%).

Solutions and Hazards

Solutions

When a solute dissolves in a solvent (or liquid) a **solution** is formed. A solution is a mixture, because the solute has not reacted with the solvent when it dissolves.

- A substance which dissolves in a solvent is **soluble**.
- A substance which does not dissolve is **insoluble**.
- When no more of the substance will dissolve in the solvent, a **saturated solution** has been formed.
- A **dilute** solution contains a lower concentration of solute than a concentrated solution.
- A **concentrated** solution can be diluted by adding water.

Top Tip
You must know the meanings of the words on the left which are in bold type.

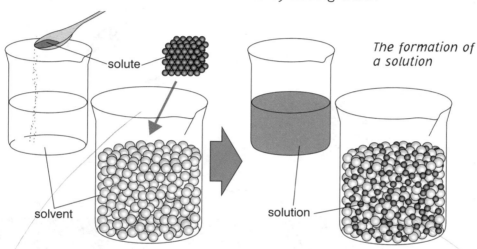

The formation of a solution

solute

solvent

solution

The fizz in fizzy drinks is due to carbon dioxide dissolved in drinks.

The label 'carbonated water' which you can see on different cans and bottles means the contents include carbon dioxide dissolved in water.

If you were to collect the gas from fizzy drinks and pass this gas through lime water, the lime water would turn cloudy or milky. Carbon dioxide is the only gas which does this to lime water and so this is the test for carbon dioxide.

Other common solutions:

- Chlorine is dissolved in water to kill bacteria.
  Chlorine is present in the drinking water from our taps.
  Chlorine is also present in water in a swimming pool but in a more concentrated solution. Sometimes it can make your eyes sore.

- Sodium fluoride is sometimes dissolved in drinking water to help prevent tooth decay.
- Lead may be dissolved in our drinking water. This happens when water comes into our homes through old lead pipes. Any lead which has dissolved in the water can be dangerous and may lead to brain damage. Lead water pipes have mostly been replaced by copper pipes or plastic pipes, which are much less dangerous.

We can say that lead in water is a **hazard** and that drinking water containing dissolved lead is hazardous to our health.

Hazards

Hazards are dangers. Many chemicals can be dangerous and their containers should be labelled with that information. The labels that are put on bottles containing dangerous chemicals are called 'hazard warning labels', and it is important that we can recognise them and know what the different hazard symbols mean.

People who work with chemicals have to follow certain regulations. These regulations or rules are for the safety of everyone who uses chemicals at work. Remember, chemicals are not just inside chemistry labs in school but are also used in industries and in shops and even in our homes. Examples of chemicals in your house which may be dangerous are bleach, oven cleaner, disinfectants and toilet cleaner. These will all have hazard symbols on their containers.

Hazard symbols which you need to know are:

 The black cross on an orange background shows that the substance is **harmful** by inhalation or if swallowed. It may also be **irritating** to eyes and to your respiratory system

 The skull and crossbones shows a much higher level of danger than "harmful", and denotes substances that are **toxic** by inhalation, **toxic** in contact with the skin and **toxic** if swallowed.

 This symbol warns that substances are **corrosive**. These chemicals can cause burns to the skin or eyes. If breathed in, they will burn the lining of the nose, throat and lungs.

 This symbol is on containers of substances that are **flammable**. The vapours of these chemicals will catch fire in the presence of a spark or a flame.

Hazard symbols are also on road tankers to indicate dangers if there is an accident and the contents spill out.

This is a symbol you might see on a tanker carrying petrol. If the tanker were to be involved in an accident, it would be important that the fire fighters knew it contained petrol.

Top Tip
You have to know the meanings of the four hazard symbols above.

Quick Test

1. What do the following mean?
   a) soluble   b) saturated solution
   c) carbonated water

2. What do you add to concentrated fruit juice to dilute it?

3. Which two substances may be added to drinking water and why?

4. What does a black cross on an orange background denote?

5. Why are corrosive chemicals dangerous?

6. What hazard symbol should be on a bottle containing a poisonous liquid?

Answers 1. a) The substance dissolves (in water); **b)** A solution in which no more of the substance can dissolve; **c)** Water containing dissolved carbon dioxide. **2.** Water. **3.** Chlorine to kill bacteria and sodium fluoride to prevent tooth decay. **4.** Harmful/irritating. **5.** These substances cause burns to the skin and eyes. **6.** The skull and crossbones on an orange background, which tells you that the substance is toxic or very poisonous.

Identifying chemical reactions

What to look for in a chemical reaction

Three things that indicate a chemical reaction has taken place include:

- a colour change

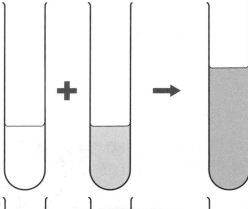

- a precipitate forming

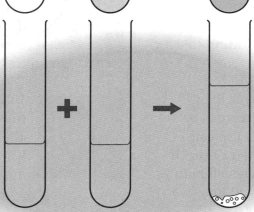

- a gas being given off

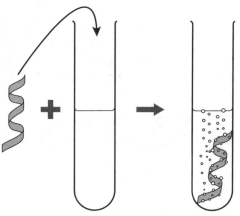

Top Tip
Remember that a precipitate is the solid formed when two liquids are mixed together.

Many different chemical reactions occur around us and inside us everyday. For example, when we are cooking and baking, chemical reactions are taking place. Other chemical reactions take place when we are digesting food.

In every chemical reaction a new substance, not present at the start, is formed. So changes in which new substances are made are chemical reactions.

Some everyday chemical reactions include:

- Boiling an egg
- Baking cakes
- Wool growing on sheep

Identifying chemical reactions by energy changes

Sometimes you cannot see any change when a chemical reaction takes place but there may be an energy change such as a change in temperature.

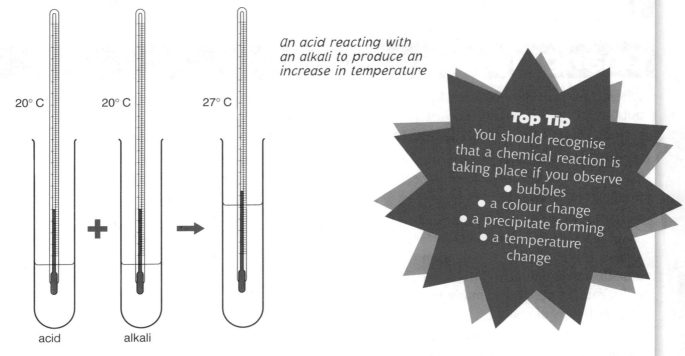

An acid reacting with an alkali to produce an increase in temperature

Top Tip
You should recognise that a chemical reaction is taking place if you observe
• bubbles
• a colour change
• a precipitate forming
• a temperature change

Some chemical reactions can be identified by temperature changes. Sometimes heat is given out and the temperature increases. In other reactions heat may be taken in and the temperature drops.

Quick Test

1. Which one of the following is **not** a chemical reaction?
   a) Iron rusting
   b) Petrol burning
   c) Ice melting
   d) Magnesium reacting with dilute acid

2. Which of the following changes are chemical reactions?
   a) Boiling water
   b) Breaking glass
   c) Frying an egg
   d) Green universal indicator turning red when placed in an acid
   e) Adding water to orange juice
   f) Separating alcohol from water by distillation
   g) Filtering sand from water
   h) Digesting our lunch
   i) Dynamite exploding

Speed of reactions

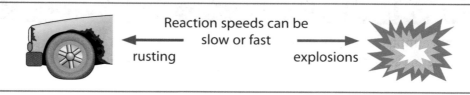

Reaction speeds can be slow or fast

← rusting

explosions →

Changing the particle size

When chalk is added to dilute hydrochloric acid, a chemical reaction takes place and bubbles of carbon dioxide gas are produced. The number of bubbles of gas produced is an indication of the reaction speed. The more bubbles seen, then the faster the reaction.

When 2 grams of large chalk lumps and 2 grams of smaller chalk lumps are added to $10\,cm^3$ of hydrochloric acid of the same concentration, the results obtained are shown in the diagrams opposite.

As you can see, the small lumps of chalk react much faster than the larger lumps. The gas given off is carbon dioxide and it will turn lime water milky much faster when made using smaller lumps of chalk.

In general, it is true that **the smaller the particle size then the faster is the chemical reaction**. Powdered chalk will react even faster than small lumps because the grains of powder are even smaller.

bubbles of gas produced slowly

bubbles of gas produced more quickly

hydrochloric acid

large lumps of chalk

small lumps of chalk

Changing the temperature

When zinc reacts with dilute hydrochloric acid, hydrogen gas is produced. Again the number of bubbles of gas gives an indication of the speed of the reaction.

Increasing the temperature always speeds up a chemical reaction. If you need to slow down a chemical reaction, such as food rotting, then the temperature should be low and the food should be kept in a fridge or freezer.

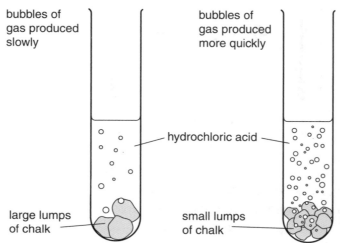

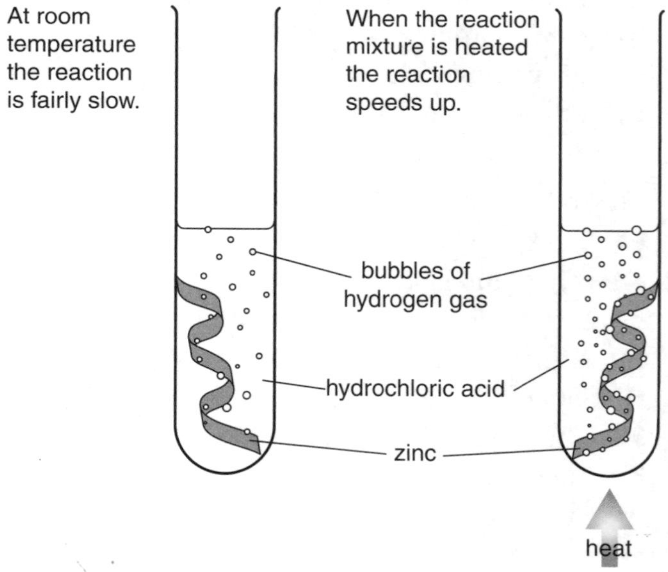

At room temperature the reaction is fairly slow.

When the reaction mixture is heated the reaction speeds up.

bubbles of hydrogen gas

hydrochloric acid

zinc

heat

Changing the concentration

Another way of speeding up the reaction between zinc and dilute hydrochloric acid is by using acid which is more concentrated.

Increasing the concentration of the reactants speeds up a chemical reaction.

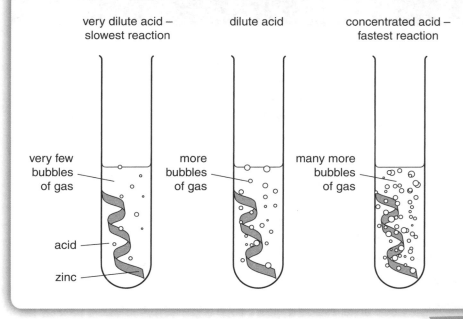

very dilute acid – slowest reaction

dilute acid

concentrated acid – fastest reaction

very few bubbles of gas

more bubbles of gas

many more bubbles of gas

acid

zinc

Top Tip

Three ways in which a chemical reaction can be speeded up are:
- Decrease the particle size
- Increase the temperature
- Increase the concentration

Quick Test

1. Write down three ways of speeding up a chemical reaction.

2. Magnesium ribbon reacts slowly with very dilute sulphuric acid at room temperature. How could this reaction be speeded up?

3. Why does milk stay fresh longer when it is kept in a fridge?

4. Rusting is a chemical reaction. Which would rust more slowly in damp conditions, iron filings or a large lump of iron?

5. Concentration is measured in moles per litre. Which would react faster with chalk, 4·0 moles per litre hydrochloric acid or 0·5 moles per litre hydrochloric acid?

Answers 1. Decrease the particle size, increase the temperature and increase the concentration. **2.** Break the magnesium ribbon into smaller pieces or use powdered magnesium; use more concentrated sulphuric acid; heat the reaction mixture gently. **3.** Milk turning sour is a chemical reaction. Decreasing the temperature will slow down this chemical reaction and the milk will stay fresh for longer. **4.** The large lump of iron. **5.** 4.0 moles per litre hydrochloric acid.

Speed of reactions and word equations

Catalysts

A catalyst is a substance which speeds up a chemical reaction. At the end of the reaction the catalyst is the same as it was at the start, i.e. a catalyst is not used up by the chemical reaction. This is very useful, since it means that the catalyst can be used over and over again.

Hydrogen peroxide is a colourless liquid which slowly decomposes to form water and oxygen gas. If enough oxygen gas is formed, it will relight a glowing splint. You will remember that this is the test for oxygen, since it is the only gas which does this.

Manganese dioxide is a black powder and when added to hydrogen peroxide it speeds up the reaction and oxygen gas is produced quickly enough for it to relight a glowing splint.

glowing splint unaffected

Glowing splint will now relight since enough oxygen gas has been produced with the catalyst present.

very, very few bubbles of oxygen gas

many more bubbles of gas

hydrogen peroxide

black powdered manganese dioxide present and acting as the catalyst

no manganese dioxide present

At the end of the reaction, the manganese dioxide is still the same as it was at the beginning. Since the manganese dioxide speeded up the reaction but was not changed by the reaction we can conclude that the manganese dioxide was acting as a catalyst.

Everyday uses of catalysts include

- Nickel in the manufacture of margarine
- Iron in the manufacture of ammonia which is used to make fertilisers
- Platinum in the manufacture of nitric acid
- Platinum in catalytic converters inside car exhaust systems. These convert poisonous carbon monoxide and oxides of nitrogen into less harmful nitrogen and carbon dioxide.

Top Tip
You should now know four ways of speeding up a chemical reaction. These are:
- Decreasing the particle size
- Increasing the temperature
- Increasing the concentration
- Using a catalyst

Catalysts in living things – enzymes

Catalysts which speed up chemical reactions in living things such as plants and animals are given the special name **enzymes**. For example, we have enzymes in our bodies which speed up the digestion of food.

Enzymes are present in biological washing powders to help remove stains such as blood, grass, food etc.

Alcoholic drinks such as beer and wine are made by a chemical reaction known as fermentation. An enzyme present in yeast speeds up fermentation.

An enzyme, catalase, which is present in our body catalyses the breakdown of hydrogen peroxide into water and oxygen. Adding a drop of blood or a piece of liver to hydrogen peroxide immediately produces enough oxygen gas to relight a glowing splint.

Word equations

The substances at the start of a chemical reaction are known as the **reactants**.

The new substances formed in a chemical reaction are known as the **products**.

In a chemical reaction the reactants change into products. This can be shown as

   Reactants → Products (The arrow means 'changes into')

Chemical reactions can be described using word equations in which the names of the reactants are written at the left-hand side of the arrow and the names of the products are written at the right-hand side of the arrow. Names of catalysts and/or enzymes are not usually written in the word equation since catalysts and enzymes do not change during the reaction.

When zinc metal reacts with dilute hydrochloric acid, zinc chloride and hydrogen gas are formed.

The names of the reactants are zinc and hydrochloric acid and so the start of the word equation is:

   zinc + hydrochloric acid →

The names of the products of this reaction are zinc chloride and hydrogen and so the completed word equation is:

   zinc + hydrochloric acid → zinc chloride + hydrogen

Top Tip
You must be able to write word equations from descriptions of chemical reactions in which you are given the names of the reactants and products.

Quick Test

1. What does a catalyst do to a chemical reaction?

2. A catalyst is unchanged at the end of a chemical reaction. What advantage does this have in the chemical industry?

3. What are biological catalysts known as?

4. Name a chemical reaction speeded up by a biological catalyst.

5. 1 gram of manganese dioxide is added to hydrogen peroxide and this catalyses the breakdown of the hydrogen peroxide. What mass of manganese dioxide will be left at the end of the reaction?

6. What is the name of the catalyst inside catalytic converters in car exhaust systems?

7. Write word equations from the information about the following chemical reactions:
   a) When magnesium burns in oxygen the product is magnesium oxide.
   b) When magnesium reacts with sulphuric acid the products are magnesium sulphate and hydrogen.
   c) In the presence of a catalyst, hydrogen peroxide breaks down into water and oxygen.
   d) Water is produced when hydrogen reacts with oxygen.
   e) Petrol burns in oxygen producing carbon dioxide and water.

Molecules, ions and formulae

Atoms

As you know, everything is made from about 100 different elements. Each element is made up of very small particles which are called **atoms**. Atoms are very, very tiny particles. A 200 cm³ glass of water contains approximately 20,000,000,000,000,000,000,000,000 atoms.

Atoms of different elements are different. The smallest atoms belong to the element hydrogen (element number 1 in the Periodic Table). Atoms of gold are about 200 times heavier than hydrogen atoms. The position of the element in the Periodic Table gives some idea of the size of the atoms of that element.

Molecules

Some substances are made up of molecules. Molecules are made up of two or more atoms joined together. Usually molecules contain atoms of non-metal elements joined together. The joins inside a molecule which hold the atoms together are called **bonds**. The bonds inside molecules are very strong.

On the other hand the bonds between different molecules are very weak. This means that molecules are easily moved apart from each other. Substances which are made up of molecules usually have low melting points and boiling points. At room temperature they are usually gases, or are liquids which turn into gases very easily.

When molecular substances are boiled, only the weak bonds between the molecules are broken, not the strong bonds inside the molecules. At low temperatures when these substances are in the solid or liquid state the molecules are very close together, but when in the gas state the molecules are far apart. This is because the bonds between the molecules have all been broken in the gas state.

In this diagram an atom is represented as ⬤ and a molecule as ⬤—⬤ .

The line — represents the strong bond inside the molecule. The bond holds the atoms together. The dotted lines ···· represent the weak bonds between the different molecules.

Since molecules do not have charges, molecular substances do not conduct electricity.

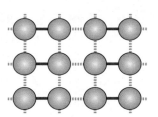

This diagram represents molecules of a substance in the solid state.

This diagram represents molecules of a substance in the gas state.

Ions

Some substances are made up of ions. Ions are tiny particles like atoms but ions are either positively charged or negatively charged. Usually ions of metal elements are positively charged and ions of non-metal elements are negatively charged.

The positive and negative ions are strongly attracted to each other and ionic compounds contain positive and negative ions bonded to each other. The bonds between oppositely charged ions are very strong. This explains why ionic compounds usually have very high melting and boiling points.

Because ions have charges, ionic compounds will conduct electricity when the ions are free to move. This happens when ionic compounds are dissolved in water and when they are melted. Most ionic compounds are soluble in water. This is shown in the table at the foot of page 4 in the Data Booklet.

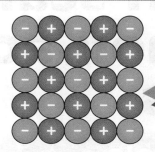

Top Tip
A table on page 4 of the Data Booklet has melting and boiling points of selected ionic compounds **and** some molecular compounds.

Formulae: using pictures and diagrams

The chemical formula of a substance tells us the number of atoms of each element present in a molecule of that substance. You probably know that the formula for water is H_2O. This tells us that there are 2 atoms of hydrogen and 1 atom of oxygen in a molecule of water. (The number '1' is never put into the formula. If no number is given then it is assumed that there is one atom of that element present.)

A molecule of water can be represented by

A molecule of propane can be represented as

$$H-\overset{\displaystyle H}{\underset{\displaystyle H}{C}}-\overset{\displaystyle H}{\underset{\displaystyle H}{C}}-\overset{\displaystyle H}{\underset{\displaystyle H}{C}}-H$$

Since there are three carbons and eight hydrogens, the formula for propane can be written as C_3H_8.

Formulae: using prefixes

Sometimes the formula of a substance can be worked out from its name. For example, the formula for carbon dioxide is CO_2. We can work this out because the prefix 'di' means 'two' and this tells us that there are two oxygens for every one carbon. The four prefixes that you must know are given in the table.

Prefix	Meaning
mono	1
di	2
tri	3
tetra	4

Top Tip
You must know the meanings of the four prefixes **mono-, di-, tri-, tetra-**

So the formula for carbon **mon**oxide is CO
sulphur **tri**oxide is SO_3
carbon **tetra**chloride is CCl_4

Quick Test

1. Why do molecular substances have low boiling points?

2. Why do ionic substances have high melting points?

3. Write the formula for:
   a) sulphur dioxide b) phosphorus trichloride
   c) nitrogen monoxide.

4. A substance has formula NO_2. What will be the name of this substance?

5. A molecule of ethanol can be represented as

$$H-\overset{\displaystyle H}{\underset{\displaystyle H}{C}}-\overset{\displaystyle H}{\underset{\displaystyle H}{C}}-O-H$$

Write down the formula for ethanol.

The pH scale and common acids and alkalis

The pH scale

The pH scale is used to work out whether a substance is neutral or acidic or alkaline.

The pH scale runs from below 0 to above 14, but most common solutions have a pH between 1 and 14.

- Acids have pH values below 7
- Alkalis have pH values above 7
- Neutral substances such as pure water have pH = 7

The pH of a solution can be measured using universal indicator, pH paper or a pH meter.

Universal indicator is red in acid, green when it is neutral and blue/purple in alkali.

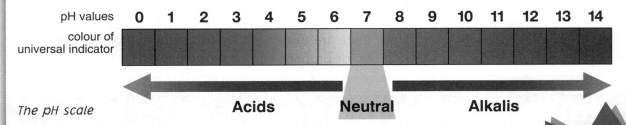

The pH scale

pH paper turns the same colours as universal indicator. Note that pH paper can only be used with liquids or solutions. When testing the pH of a solid substance the pH paper must be moistened first or the substance dissolved to make a solution.

A pH meter gives a read-out of the actual pH of the solution. It is more sensitive than universal indicator or pH meter and usually gives the pH value to one decimal place.

Top Tip
Remember that the lower the pH value then the more acidic the solution. The higher the pH then the more alkaline the solution.

Diluting acids and alkalis

When an acid is diluted by adding water, its acidity decreases. Likewise, adding water to an alkali decreases its alkalinity (makes it less alkaline).

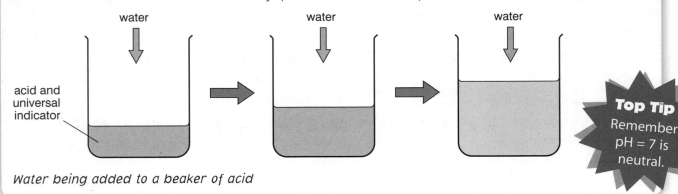

Water being added to a beaker of acid

Top Tip
Remember pH = 7 is neutral.

Common acids and alkalis

We use acids and alkalis regularly at home as well as in the school laboratory.

Both acids and alkalis are corrosive and should be used carefully.

Alkalis are often more corrosive than acids.

Common laboratory acids are:	Common laboratory alkalis are:
• hydrochloric acid • sulphuric acid • nitric acid	• sodium hydroxide • lime water • ammonia
Common household acids are: • vinegar • lemonade • soda water • cola drinks • fruit juices	Common household alkalis are: • baking soda • oven cleaner • dishwashing powder • bleach

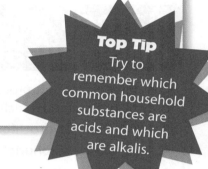

Top Tip
Try to remember which common household substances are acids and which are alkalis.

Quick Test

1. A solution has pH = 3. Is it acid, alkaline or neutral?
2. What three things can be used to test the pH of a solution?
3. What colour does pH paper turn in alkali?
4. What colour is universal indicator in a solution of pH = 4?
5. What colour is universal indicator in pure water and why?
6. What happens to the pH of an acid when it is diluted with water?
7. What happens to the pH of an alkali when it is diluted with water?
8. Name three laboratory acids.
9. Name three laboratory alkalis.
10. Is vinegar an acid or an alkali?
11. Is oven cleaner an acid or an alkali?

Answers 1. Acid. **2.** Universal indicator, pH paper and a pH meter. **3.** Blue-purple. **4.** Orange. **5.** Green, because water is neutral. **6.** It rises towards pH = 7. **7.** It falls to pH = 7. **8.** Hydrochloric, sulphuric and nitric acids. **9.** Sodium hydroxide, lime water and ammonia. **10.** Acid. **11.** Alkali.

21

Neutralisation and acid rain

What is neutralisation?

- A neutralisation reaction is one in which a substance cancels out the effect of an acid or alkali.
- Acids react with alkalis in a neutralisation reaction.
- In a neutralisation reaction the pH of an acid rises towards 7.
- In a neutralisation reaction the pH of an alkali falls towards 7.
- In a neutralisation reaction, water and a salt are always formed.

Names of salts formed during neutralisation

The name of the salt produced during a neutralisation reaction depends on which alkali and which acid reacted together.

- Neutralising **hydrochloric** acid produces **chloride** salts.
- Neutralising **sulphuric** acid produces **sulphate** salts.
- Neutralising **nitric** acid produces **nitrate** salts.

Top Tip
You need to be able to work out the names of the salts produced during a neutralisation reaction.

For example:

When the alkali **sodium** hydroxide reacts with **hydrochloric** acid, the salt formed is called **sodium chloride**.

When the alkali **sodium** hydroxide reacts with **sulphuric** acid, the salt formed is called **sodium sulphate**.

When the alkali **calcium** hydroxide reacts with **nitric** acid, the salt formed is called **calcium nitrate**.

Metal carbonates neutralising acids

It is not just alkalis which will neutralise acids. Metal carbonates also neutralise acids. This time the products are not just water and a salt. Carbon dioxide gas is also formed in the reaction. The carbon dioxide formed turns lime water milky.

When **magnesium** carbonate reacts with **hydrochloric** acid the name of the salt is **magnesium chloride**.

Carbon dioxide is produced when a metal carbonate reacts with acid.

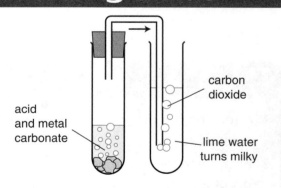

acid and metal carbonate

carbon dioxide

lime water turns milky

Everyday examples of neutralisation

- Putting vinegar onto the alkaline sting from a wasp.
- Spreading lime on fields in which the soil is too acidic.
- Taking antacid tablets to treat acid indigestion in your stomach.
- Using lime (or limestone) to reduce the effects of acid rain on lakes.

Vinegar can be used to treat a wasp sting.

The gases which cause acid rain

The gases carbon dioxide (CO_2), sulphur dioxide (SO_2) and nitrogen dioxide (NO_2) all dissolve in water to form acidic solutions.

Carbon dioxide is formed when carbon burns in oxygen or air.

Although carbon dioxide is an acidic gas it is not thought to cause acid rain.

The acidic gases which are thought to be mainly responsible for acid rain are sulphur dioxide and nitrogen dioxide.

Sulphur dioxide can be made by burning sulphur in air or oxygen. Most sulphur dioxide is formed when fossil fuels are burned. This is because fossil fuels such as coal and oil contain sulphur or sulphur compounds in tiny quantities.

The nitrogen and oxygen present in air will react together to form nitrogen dioxide under extreme conditions such as during lightning storms or around the very hot sparks in petrol engines.

When **sulphur dioxide** and **nitrogen dioxide** dissolve in water vapour and clouds, the rain that falls can have a pH below 5 and this is known as **acid rain**.

The damage caused by acid rain

- Acid rain damages buildings and statues made of marble and limestone.
- Acids corrode iron and steel and so acid rain will cause serious damage to structures such as bridges made from iron.
- Acid rain can also pollute soils to the extent that plants will not grow properly, or not grow at all.
- Acid rain may also pollute lochs, lakes and ponds to the extent that fish, plants and other living things will not survive.

Top Tip
Sulphur dioxide and nitrogen dioxide are the gases mainly responsible for acid rain. You should be able to give examples of the damage caused by acid rain.

Quick Test

1. What happens to the pH of an acid when it is neutralised?

2. What is the name of the salt formed when calcium hydroxide neutralises hydrochloric acid?

3. Which acid would be used to prepare potassium sulphate?

4. Which gas is produced when calcium carbonate reacts with sulphuric acid, and what is the test for this gas?

5. What is added to soil to reduce soil acidity?

6. What might be put on a wasp sting to reduce the pain?

7. Which two gases are thought to be the main causes of acid rain?

Answers 1. It rises towards 7. **2.** Calcium chloride. **3.** Sulphuric acid. **4.** Carbon dioxide, which turns lime water milky. **5.** Lime (or limestone). **6.** Vinegar. **7.** Sulphur dioxide and nitrogen dioxide.

PPA 1: The effect of temperature on speed of dissolving

Introduction

When sugar crystals dissolve in water they form a solution. The speed at which the sugar dissolves depends on various factors. In this PPA the factor that was changed was the temperature of the water.

Aim

To find the effect of changing the temperature of the water on the speed at which sugar crystals dissolve.

Procedure

- Using a syringe you filled a test tube with water to about 3 cm from the top.
- You used a thermometer to measure the initial temperature of the water.
- One spatula of sugar was added to the test tube and you then stoppered the test tube.
- You then turned the test tube upside down and held it there until the sugar crystals had fallen to the bottom. This was counted as one **upturn**.
- The test tube was then turned the right way up until the crystals again fell to the bottom. This was counted as the **second upturn**.
- You repeated this until all the sugar crystals had disappeared or dissolved, counting the total number of upturns required. The number of upturns gives an indication of how quickly the sugar dissolves.
- As soon as the sugar had dissolved you removed the stopper from the test tube and measured the final temperature of the water.
- You were then able to calculate the average temperature of the water from the initial and final temperature values.
- You then used a Bunsen burner to heat some more water until the temperature was between 35 °C and 40 °C and repeated the experiment using this warmer water.
- You then heated more water, being careful not to go beyond 60 °C and repeated the experiment using this hotter water.

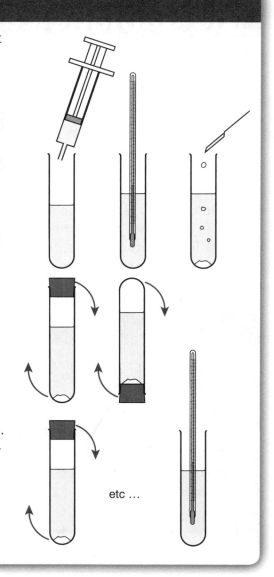

etc ...

Results

Results similar to ones you may have got are shown in the table below.

Initial temperature of water in °C	Final temperature of water in °C	Average temperature in °C	Number of 'upturns'
22	22	22	19
38	34	36	12
56	50	53	7

Conclusion

From the table of results you can see that as the temperature of the water increases the number of upturns required decreases. We can conclude that as the temperature increases the speed at which sugar dissolves also increases.

Points to note

- In this PPA, the aim was to find out how changing the temperature affected the speed at which sugar dissolves, so the factor changed was the temperature of the water.
- The number of 'upturns' required indicated how quickly the sugar dissolved.
- The higher the temperature, the faster the sugar dissolved.

In many ways the experiment was not fair. Ways that it might have been improved include:
- Measuring the quantity of sugar used more accurately than just using a spatula
- Measuring the volume of water more accurately each time
- Doing the upturns at the same speed every time

Quick Test

1. What did you count that let you know how quickly the sugar dissolved in the water?
2. Why was the temperature measured at the start and at the end?
3. What did you find out from the results of the experiment?
4. Write down any way in which the experiment could have been improved.

Answers 1. The number of upturns. 2. So that you could calculate the average temperature for each experiment. 3. The higher the temperature, the faster the sugar dissolved. 4. The quantity of sugar used each time could have been measured more accurately so that it was exactly the same each time.

PPA 2: The effect of concentration on reaction speed

Introduction

When a piece of magnesium reacts with sulphuric acid, bubbles of hydrogen gas are produced. The time taken for the piece of magnesium to slowly disappear gives some idea of the reaction speed. The **longer** the time taken the **slower** the reaction speed. The speed of the reaction depends on various factors, and the one changed in this PPA was concentration.

Aim

The aim of this experiment was to find out how changing the concentration affected the speed of reaction between sulphuric acid and magnesium ribbon.

Procedure

- You used a syringe to measure out 20 cm^3 of sulphuric acid into a glass beaker. The acid had a concentration of 2 moles per litre.
- A 2 cm piece of magnesium ribbon was added to the acid and a timer was started.
- The time taken for the magnesium to disappear was measured and recorded.
- The glass beaker was then washed and dried.
- Sulphuric acid of concentration 1 mole per litre was prepared by adding 10 cm^3 of the 2 moles per litre acid to 10 cm^3 water in the glass beaker.
- The experiment was repeated using the sulphuric acid of concentration 1 mole per litre and the time taken for the magnesium to disappear was again noted.
- Sulphuric acid of concentration 0·5 moles per litre sulphuric acid was then prepared by adding 5 cm^3 of the 2 moles per litre acid to 15 cm^3 of water in the glass beaker.
- The experiment was repeated using the sulphuric acid of concentration 0·5 moles per litre and the time taken for the magnesium to disappear was again noted.

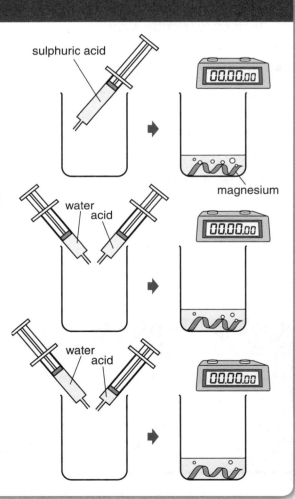

sulphuric acid

magnesium

water
acid

water
acid

Results

Results similar to ones you may have got are shown in the table below.

Concentration of acid in moles per litre	Time for magnesium to disappear in seconds
2	10
1	18
0·5	38

Conclusion

From the table of results you can see that as the concentration of the acid decreases the time for the magnesium to disappear increases. We can conclude that as the concentration increases, reaction speed also increases.

Points to note

- In this PPA, the aim was to find out how changing the concentration of the acid affected the speed at which it reacts with magnesium so the only factor changed was the concentration of the acid.
- The concentration of the acid was changed by adding it to water.
- Factors which had to be kept the same in the experiments were:
  - temperature
  - length of magnesium ribbon
  - volume of acid
  - particle size of the magnesium (always ribbon, not powder or lump)
- The time taken for the magnesium to disappear gave a measure of the reaction speed.
- In each experiment it was important that the magnesium was submerged in the acid. This is because the reaction time would have been much longer if the magnesium ribbon had just floated on the surface of the acid.
- The results of this PPA showed that increasing the concentration increases the reaction speed.

Quick Test

1. Which gas is produced in the experiment?

2. How was the concentration of the acid changed from 2 moles per litre to 1 mole per litre?

3. When did you stop the timer?

4. What happened to the reaction speed when the acid was diluted?

5. What factors had to be kept the same for each experiment?

6. Why did the magnesium ribbon have to be submerged in the acid?

Answers 1. Hydrogen gas. **2.** 10cm³ of the 2 moles per litre acid was added to 10cm³ of water. **3.** When all the magnesium had disappeared. **4.** The reaction speed became slower. **5.** Temperature, length of magnesium ribbon, volume of acid, particle size of the magnesium. **6.** If the magnesium had not been submerged but had floated on the surface of the acid there would not have been such good contact between the magnesium and the acid, and the reaction would have been slower than it should have been.

PPA 3: Testing the pH of solutions

Introduction

The pH scale is used to measure how acidic or alkaline a solution is. The pH scale is from just below 0 to just above 14. The pH of a solution can be measured either by using pH indicator or pH paper.

When either pH indicator solution or pH paper is put into an acid or an alkali it changes colour. You can match this colour against the colours on the pH colour chart and read off the pH of the solution.

Acidic solutions have pH below 7 and turn pH indicator and pH paper red/orange.

Alkaline solutions have pH above 7 and turn pH indicator and pH paper blue/purple.

Neutral solutions have pH = 7 and turn pH indicator and pH paper green.

In this PPA you had a choice whether you used indicator solution or pH paper.

Using pH indicator solution

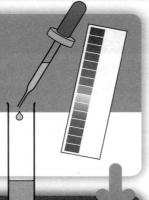

Using pH paper

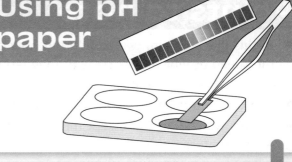

Aim

To use pH indicator to find the pH of some household substances and to decide whether they are acidic, alkaline or neutral.

Aim

To use pH paper to find the pH of some household substances and to decide whether they are acidic, alkaline or neutral.

Procedure

- You added about 2 cm of vinegar to a test tube and then added 2 or 3 drops of the pH indicator solution and shook the test tube.
- The indicator changed colour and to get the pH you matched the colour of the solution to one on the pH colour chart.
- This procedure was repeated with the other household solutions.
- If the substance was a solid you had to dissolve it in a small amount of water before adding the pH indicator solution.

Procedure

- You added a few drops of vinegar to a dimple tile and then used tweezers to dip a piece of pH paper into the vinegar.
- The pH paper changed colour and to get the pH you matched the colour of the paper to one on the pH colour chart.
- This procedure was repeated with the other household solutions.
- If the substance was a solid you had to dissolve it in a small amount of water before putting it into the dimple tile.

Results

The results should have been the same no matter which method you used and should have been similar to the ones in the table below.

Household substance	pH	Acidic/alkaline/neutral
Vinegar	3	Acidic
Soda water	4	Acidic
Salt solution	7	Neutral
Lemon juice	5	Acidic
Diluted ammonia	11	Alkaline
Bicarbonate of soda	11	Alkaline
Automatic washing powder	9	Alkaline
Sugar	7	Neutral

Conclusion

- Vinegar, soda water and lemon juice are acidic.
- Salt and sugar solutions are neutral.
- Ammonia, bicarbonate of soda and washing powder solutions are alkalis.

Points to note

In this PPA, the aim was to use pH indicator solution or pH paper to find the pH of some household substances and to decide whether they were acidic, alkaline or neutral.

- A solution which has pH = 7 is neutral.
- A solution which has pH below 7 is acidic.
- A solution which has pH above 7 is alkaline.
- To get the pH value you have to match the colour of the pH indicator or pH paper with the colour chart and read off the pH value.
- If the substance is a solid it must firstly be dissolved in water.
- If the substance is a solid which does not dissolve in water then it is impossible to measure its pH.

Quick Test

1. What is the pH of a neutral substance?

2. What 2 methods can be used to measure the pH of a solution of a substance?

3. Why is it necessary to use the pH colour chart?

4. What colour does pH paper turn in an alkaline solution?

5. If the substance is a solid what do you have to do first before measuring its pH?

6. Why is it impossible to measure the pH of certain substances such as sand, glass and marble?

Answers 1. pH = 7. **2.** Using pH paper or pH indicator solution. **3.** To match up the colour that the pH paper or pH indicator change into and so read off the pH value. **4.** Blue/purple. **5.** Dissolve it in water first. **6.** Sand, glass and marble are insoluble solids.

Uses of metals

Extraction of metals

Only a few metals, such as **gold**, **silver** and **copper**, are found in the Earth's crust in an **uncombined state**. This means that they are present as elements and not as compounds. The vast majority of metals, however, are found in the **combined state**. This means they are present as compounds. Naturally occurring compounds of metals are known as **ores**.

Before metals can be used, they have to be extracted from their ores. The method used to do this depends on the reactivity of the metal. If you look at page 6 of your Data Booklet, you will see some common metals listed in order of reactivity. Reactive metals, like **aluminium**, are extracted from their ores by using **electricity**. The reason is that a lot of energy is needed to change these ores into metals. For less reactive metals, less energy is needed to extract them from their ores. Iron, for example, is extracted from its ore by heating it with carbon.

Top Tip
Remember that aluminium is extracted from its ore by using electricity and that iron is extracted by heating its ore with carbon.

Properties of metals

Some of the important properties of metals are listed below:

- **Strength**
  Metals are usually very **strong** and this is why iron, for example, is used to make bridges, buildings, cars, trains and ships.

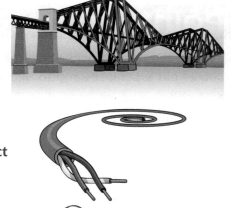

- **Conductors of electricity**
  Metals are good **conductors of electricity** and this is why copper, for example, is used in electric cables.
  It is important to remember that **all metal elements conduct electricity**. **Non-metal elements**, apart from carbon in the form of graphite, **do not conduct electricity**.

- **Conductors of heat**
  Metals are good **conductors of heat**. This is why aluminium, for example, can be used to make sauce pans.

- **Malleable**
  Metals are **malleable**. This means that they can be beaten into different shapes. Gold, for example, can be beaten into sheets so thin that it can be used in the protective visors of astronauts' space helmets.

- **Density**
  If you look at page 3 of your Data Booklet, you will see that aluminium has a density of 2·70 units. This means that 1 cm³ of aluminium weighs 2·70 grams. Iron, on the other hand, has a density of 7·87 units, i.e. 1 cm³ of iron weighs 7·87 grams. For the same volume, aluminium is lighter than iron and explains why aluminium, rather than iron, is used in making aircraft.

Alloys

An **alloy** is **a mixture of metals** or **a mixture of metals and non-metals**. The reason why a metal is mixed or alloyed with another element is to change its properties and make it more suitable for particular uses. For example, iron is mixed with tiny amounts of carbon to make the alloy, steel. Steel is far stronger than iron and is a more suitable material for making bridges, buildings, cars etc.

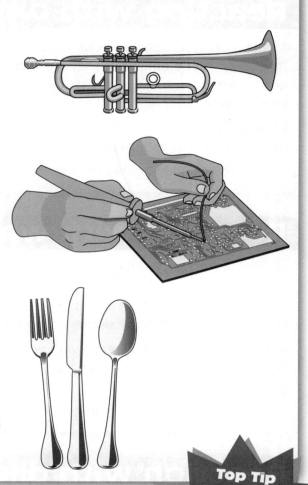

- **Brass**

  Brass is an alloy of copper and zinc and is much harder and stronger than either. Brass is used in musical instruments, ornaments, plaques and door handles.

- **Solder**

  Solder is an alloy of tin and lead and is used to make electrical connections. It can be used in this way because it has a much lower melting point than either tin or lead. When held against a hot soldering iron, the solder melts and makes the electrical connection.

- **'Stainless' steel**

  'Stainless' steel is an alloy of iron, carbon, chromium and nickel and is used in cutlery, kitchen sinks and watches. The chromium and nickel in the 'stainless' steel stops it rusting.

Top Tip
Make sure you know that an alloy is a mixture of metals or a mixture of metals and non-metals.

Quick Test

1. Name three metals which are found uncombined in the Earth's crust.

2. What is used to extract aluminium from its ore?

3. Look at page 8 of your Data Booklet and state whether each of the following elements is a conductor or a non-conductor of electricity:
   a) magnesium   **b)** bromine   **c)** graphite   **d)** mercury

4. Give a reason why aluminium is used in making aircraft.

5. What is an alloy?

Answers 1. Gold, silver and copper. **2.** Electricity. **3. a)** conductor; **b)** non-conductor; **c)** conductor; **d)** conductor. **4.** Because it has a low density or is light. **5.** It is a mixture of metals or a mixture of metals and non-metals.

Reactions of metals

Reaction with oxygen

Metals such as potassium, sodium and lithium react so readily with the oxygen in the air that they have to be stored under oil. They are very **reactive** metals. Gold and silver, on the other hand, don't have to be stored in any special way because they do not react with oxygen. They are very **unreactive** metals.

In general, when a metal reacts with oxygen, a metal oxide is formed. For example, when magnesium reacts with oxygen, magnesium oxide is produced. The word equation is:

magnesium + oxygen → magnesium oxide

Reaction with water

The Column 1 metals, lithium, potassium and sodium, react vigorously with water. The potassium/water reaction is the fastest and the lithium/water reaction is the slowest. By comparing the reaction speeds we can place these metals in order of reactivity: the faster the reaction the more reactive the metal. This means that potassium is the most reactive, then sodium and finally lithium.

In general, when a metal reacts with water, hydrogen gas and a metal hydroxide are formed. For example, sodium and water react to produce hydrogen and sodium hydroxide. The word equation is:

sodium + water → hydrogen + sodium hydroxide

Reaction with dilute acid

When a metal reacts with a dilute acid, bubbles of hydrogen gas are given off. The speed at which the bubbles are produced gives us some idea of the reactivity of the metal: the greater the speed the more reactive the metal. The experiment shown opposite can be set up to place the metals, magnesium, copper and zinc in order of reactivity.

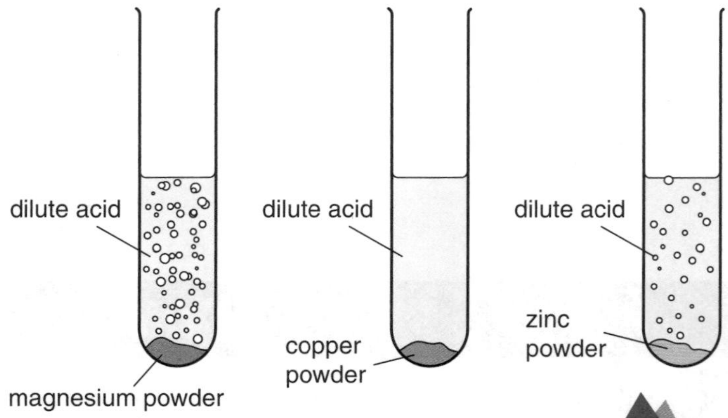

From these results, we can see that the order of reactivity, starting with the most reactive, is magnesium, zinc and then copper.

dilute acid — magnesium powder

dilute acid — copper powder

dilute acid — zinc powder

You'll notice that no bubbles of hydrogen gas are produced in the second test tube. This is because copper does not react with dilute acid.

In general, when a metal reacts with a dilute acid, hydrogen gas and a salt are formed. For example, the reaction between zinc and dilute sulphuric acid produces hydrogen and the salt, zinc sulphate. The word equation is:

zinc + dilute sulphuric acid → hydrogen + zinc sulphate

Top Tip

A metal reacts
• with oxygen to form a metal oxide
• with water to form hydrogen and a metal hydroxide
• with dilute acid to form hydrogen and a salt.

Test for hydrogen

When a lighted splint is held at the mouth of a test tube full of hydrogen gas, the hydrogen burns with a squeaky 'pop'.

This is the **test for hydrogen**.

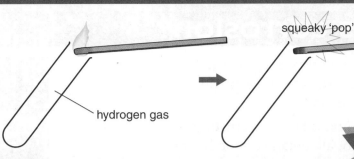

squeaky 'pop'

hydrogen gas

Top Tip
Remember the test for hydrogen – it burns with a squeaky 'pop'.

Reactivity series

By taking a wide variety of metals and looking at their reaction speeds with oxygen, water and dilute acid, a longer list of metals, arranged in order of reactivity, can be obtained. This list is shown in the first column of the table below. As we go down the column, the reactivity of the metals decreases. The reactions of metals with oxygen, water and dilute acid are also summarised in the table.

Metal	Reactivity	Reaction with oxygen	Reaction with water	Reaction with dilute acid
potassium sodium lithium calcium magnesium aluminium zinc iron tin lead copper mercury silver gold	most reactive ↓ decreasing reactivity ↓ least reactive	metals which react with oxygen to form metal oxides metals which do **not** react with oxygen	metals which react with water to form hydrogen gas and metal hydroxides metals which do **not** react with water	metals which react with dilute acid to form hydrogen gas and salts metals which do **not** react with dilute acid

Top Tip
There is a table like this on page 6 of your Data Booklet.

Quick Test

1. Write a word equation for the reaction between copper and oxygen.

2. Name the products formed when potassium reacts with water.

3. Write a word equation for the reaction between magnesium and hydrochloric acid.

4. What is the test for hydrogen?

5. Metal **A** reacts with water; metal **B** reacts with oxygen but not with dilute acid; metal **C** reacts with dilute acid but not with water.
   a) Use this information and the table above, to list the metals in order of reactivity, starting with the most reactive.
   b) Suggest a name for metal **B**.

Answers 1. copper + oxygen → copper oxide. **2.** Hydrogen and potassium hydroxide. **3.** magnesium + hydrochloric acid → hydrogen + magnesium chloride. **4.** It burns with a squeaky 'pop'. **5. a)** A, C, B; **b)** Copper.

Corrosion

What is corrosion?

Corrosion is a chemical reaction that takes place on the surface of a metal. The metal reacts with substances in the air to form a compound.

The corrosion of iron is known as **rusting** and this term only applies to iron. So, all metals corrode but only iron rusts.

Iron loses its strength when it rusts and structures made of iron, like the car opposite, are weakened and become unsafe.

Conditions needed for rusting

The experiment shown opposite can be set up to show which substances in the air cause iron to rust.

- In test tube **A**, the iron nail is in contact with both air and water.
- In test tube **B**, the nail is in contact with air but not water. The drying agent removes the water from the air.
- In test tube **C**, the nail is in contact with water but not air. The water has been boiled to remove the dissolved air and the oil layer prevents any air from re-dissolving.

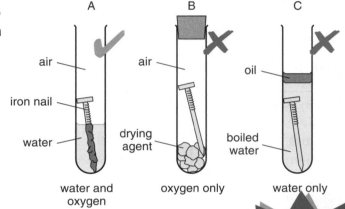

After a few days, rust forms only in test tube **A**, where the iron is in contact with both air and water. Air contains several gases, but it can be shown that when iron rusts, the gas that is used up is oxygen. So, iron rusts only when oxygen and water are both present.

Top Tip
Remember that both oxygen and water must be present for iron to rust.

Factors that affect the speed of rusting

The experiment opposite can be set up to show what effect **acid rain** and **salt water** have on the speed at which iron rusts.

Test tube **A** is the control – the iron nail is in water. In **B**, the nail is in acid rain and in **C**, the nail is in salt water.

In all three test tubes, **rust indicator** is present. This indicator turns blue if rusting occurs, and the deeper the blue colour the greater the degree of rusting.

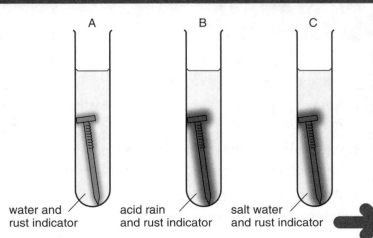

The blue colour in test tubes **B** and **C** is more intense than it is in **A** and this proves that both acid rain and salt increase the speed of rusting.

The undersides of cars rust more in winter than they do in summer. This is because of the salt that is spread on roads during the winter months.

Rust prevention

If iron is to rust, it must be in contact with oxygen and water. If a coating is put on the surface of the iron, it will act as a barrier to the oxygen and water and so prevent rusting. Examples of this type of protection include:

- **Painting**: car bodies, iron gates and bridges (for instance, the Forth Rail Bridge)
- **Oiling and greasing**: moving parts in industrial and agricultural machinery, bike chains
- **Plastic coating**: garden wires and fences, fridge shelves
- **Tin-plating**: food cans (tin cans). Traditionally, tin cans were made of iron on which there was a thin coating of tin.
- **Galvanising**: car exhausts and nails. Galvanised iron is iron coated with zinc. This is done by dipping the iron into molten zinc.
- **Electroplating**: silver-plated cutlery, chromium-plated car bumpers. Electroplated iron has a thin layer of another metal coated on it. This can be achieved by electrolysis.

Another way of preventing iron from rusting is to attach a more reactive metal to the iron structure. Zinc is more reactive than iron (see page 6 of your Data Booklet), and this is why zinc blocks are bolted onto the hulls of ships and the legs of oil rigs to stop them rusting. Scrap magnesium is often attached to underground iron pipes and tanks to stop rusting. It can do this because magnesium is more reactive than iron.

Top Tip
Iron will not rust if it is
- completely coated by another material
- attached to a more reactive metal.

Galvanised iron versus tin-plated iron

We have just learned that the zinc coating on galvanised iron and the tin coating on tin-plated iron prevent rusting because they stop oxygen and water from coming into contact with the iron. However, what happens if the coatings are scratched and the iron is exposed? Is the iron still protected against rusting?

The zinc will continue to prevent rusting because it is more reactive than iron. Tin, however, is less reactive than iron and will not prevent rusting. In fact, the iron will rust even more rapidly than an untreated piece of iron.

Quick Test

1. Into what type of substance is a metal element changed, when it corrodes?
2. Which two substances must be present before iron will rust?
3. Apart from acid rain, name another substance which speeds up rusting.
4. Iron which has been painted does not rust. Explain why the paint stops the iron rusting.
5. a) What name is given to iron coated with zinc?
   b) Why doesn't the iron rust even when the zinc coating is scratched?

Answers 1. A compound. **2.** Oxygen and water. **3.** Salt. **4.** The paint acts as a barrier and prevents oxygen and water reaching the iron. **5. a)** Galvanised iron; **b)** Because zinc is more reactive than iron.

Batteries

Batteries or cells?

Mobile phones, cameras, radios, calculators, remote controls, laptops and iPods are just a few of the things that are powered by batteries. Batteries come in different shapes and sizes but they all provide us with a convenient and portable source of electricity.

The electricity is produced when the chemicals in the battery react together but once the chemicals are used up, the chemical reaction stops and the battery can no longer produce electricity. At this point, the battery has to be replaced or if it is a rechargeable battery, like the car battery below, it has to be recharged.

The words 'battery' and 'cell' are often used to mean the same thing but, strictly speaking, they are different. Something which produces electricity from a chemical reaction is a cell, and a battery is a collection of two or more cells joined together.

The lead-acid or car battery

A typical car battery has six cells linked together. Each cell consists of lead plates dipping into dilute sulphuric acid, and this is why a car battery can also be described as a lead-acid battery.

Top Tip
Remember that in a battery, electricity comes from a chemical reaction.

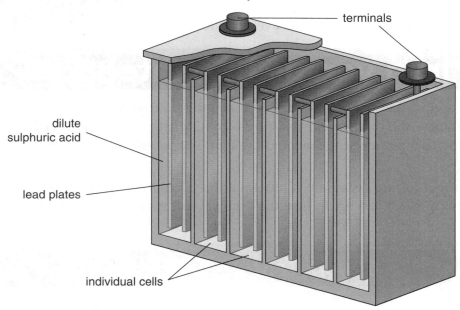

terminals

dilute sulphuric acid

lead plates

individual cells

cutaway diagram of a lead-acid battery

A car battery is rechargeable. This means that once the chemical reaction stops and the battery is no longer producing electricity, it can be recharged. In the recharging process, the chemical reaction is reversed and the chemicals needed to produce electricity in the first place are remade.

Simple cells

A simple cell, like the one shown opposite, consists of two different metals dipping into a solution containing ions.

With this arrangement, a reading is obtained on the voltmeter which proves that electricity is produced in the cell.

For electricity to be generated the solution in the cell must contain ions, and the purpose of this ion solution is to complete the circuit. A solution of common salt, i.e. sodium chloride solution, can be used as the ion solution.

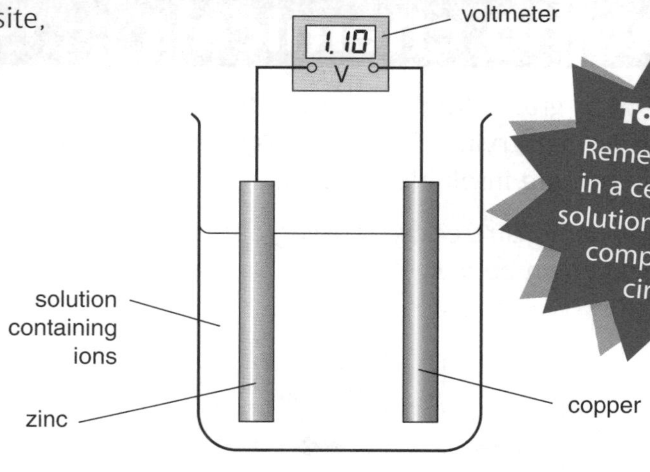

voltmeter

1.10 V

solution containing ions

zinc

copper

a simple cell

Top Tip
Remember that in a cell, the ion solution is there to complete the circuit.

The table below shows the voltages that are obtained when cells are set up with pairs of different metals.

Metal pair		Voltage in volts
magnesium	silver	3.17
zinc	copper	1.10
iron	tin	0.30
lead	lead	0.00

The first three results show that different pairs of metals produce different voltages and the last result shows that the two metals in a cell must be different if electricity is to be produced.

The size of the voltage depends on the difference in the reactivity of the two metals used in the cell. The bigger the difference in reactivity, the bigger is the voltage.

If you look at page 6 of the Data Booklet, you'll see that of the seven metals mentioned in the table above, magnesium and silver are the furthest apart in the reactivity series, i.e. they have the biggest difference in reactivity. This is why the magnesium/silver cell produces the largest voltage.

Quick Test

1. What produces the electricity in a battery?
2. Give an example of a rechargeable battery.
3. What is the purpose of the ion solution in a cell?
4. After a cell has been used for some time, its voltage decreases. Why does this happen?
5. The following metal pairs were used in cells.

   A. zinc/silver;   **B.** copper/silver;   **C.** aluminium/silver

   Look at page 6 of the Data Booklet and decide which of the above metal pairs would produce the lowest voltage.

Answers 1. A chemical reaction. **2.** A car battery or a lead-acid battery. **3.** To complete the circuit. **4.** The chemicals in the cell are being used up. **5.** B.

Keeping clean

Cleaning chemicals

Cleaning oil or grease from the skin or from clothes is almost impossible using water alone. The same is true when trying to clean greasy hair. The problem is that oil and grease don't dissolve in water, i.e. they are insoluble in water.

This can be demonstrated by carrying out the experiment shown below.

In this next experiment, a few drops of detergent have been added to the water.

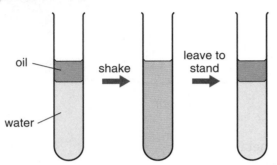

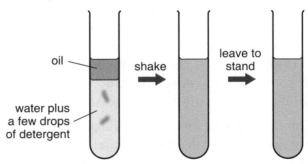

When the test tube is shaken, the oil and water mix as the oil breaks up into tiny droplets. But when the mixture is left to stand, these oil droplets join up again to form a separate layer once more.

On shaking, the oil and water mix as before but the difference is that they remain mixed when left to stand. What happens is that the detergent particles surround the oil droplets and prevent them joining up again.

So, the problem of cleaning oil or grease from skin, hair, clothes etc. is solved by adding a **cleaning chemical**, such as a detergent, to the water. The cleaning chemical keeps the grease or oil mixed with the water and this mixture can be washed away, leaving the surface free of oil or grease. Cleaning chemicals work because not only are they soluble in water, they are also soluble in oil and grease.

There are lots of products which contain cleaning chemicals. Examples include soaps, detergents, washing-up liquids, washing powders and shampoos.

Hard and soft water

In some areas of the country the water is described as **hard** while in others it is **soft**. For example, water in the Glasgow area is soft but in London the water is very hard.

When soap is added to soft water, a lather is readily formed.

However, when soap is added to hard water, little or no lather is produced. Instead, a grey solid called 'scum' is formed. It clings to clothes and leaves 'tidemarks' in baths and sinks.

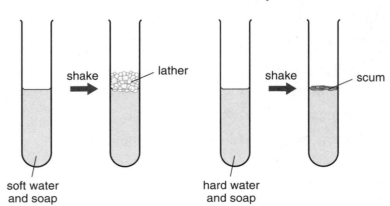

Hard water contains calcium and magnesium compounds and it is these that make it hard. They react with the soap to form the scum. As well as being unsightly, scum causes another major problem. It prevents the soap acting as a cleaning chemical. The reason for this is that the soap is locked up in the scum and is not free to clean.

Top Tip
Scum is formed when soap is added to hard water.

Scientists have devised various ways of tackling the problem of hard water. One way was to design cleaning chemicals which would not form a scum with hard water. These cleaning chemicals are known as **soapless detergents** and when they are added to hard water, lather is formed, but no scum.

Dry-cleaning

Although oil and grease are insoluble in water, they are soluble in some liquids. These liquids can be described as solvents because the oil and grease dissolve in them. It is these **special solvents** that are used in the **dry-cleaning** industry.

Dry-cleaning is normally used for clothes which would be damaged by water or detergents. Dry-cleaning machines are just like household washing machines but instead of using water, the clothes are washed in these special solvents.

A dry-cleaning solvent must:
- be good at dissolving substances like oil and grease
- be non-flammable, i.e. it must not catch fire easily
- be non-toxic so that the health of the operators is not damaged
- evaporate easily so that the clothes dry quickly
- not damage the clothes.

Top Tip
Remember that dry-cleaning uses special solvents which are good at dissolving oil and grease.

Quick Test

1. Why are greasy plates difficult to clean in water alone?

2. Which of the following statements is true?

   A cleaning chemical is
   A   soluble in water and soluble in oil
   B   soluble in water and insoluble in oil
   C   insoluble in water and soluble in oil
   D   insoluble in water and insoluble in oil.

3. Give two examples of cleaning chemicals.

4. What is formed when soap is added to hard water?

5. Apart from using washing powder and water, what other method could be used to clean oil stains from clothes?

Answers 1. Because grease is insoluble in water or does not dissolve in water; **2. A;** **3.** Any two from soaps, detergents, shampoos, washing-up liquids and washing powders. **4.** Scum. **5.** Dry-cleaning.

Clothing

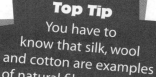

Natural and synthetic fibres

The main reason for wearing clothes is to protect our bodies from the environmental conditions in which we live e.g. extremes of heat and cold, rain and snow, strong winds, strong sunlight etc.

Clothes are made from pieces of **fabric** or cloth, and fabrics consist of a network of thin thread-like strands, known as **fibres**. There are two main types of fibre – natural and synthetic.

Natural fibres come from animals and plants and examples include:

- silk made by the silkworm
- wool from sheep
- cotton from the cotton plant

Synthetic fibres are manufactured by the chemical industry. **Nylon**, which is a polyamide, and **Terylene**, which is a polyester, are just two examples of a vast range of synthetic fibres.

> **Top Tip**
> You have to know that silk, wool and cotton are examples of natural fibres and that nylon and Terylene are examples of synthetic fibres.

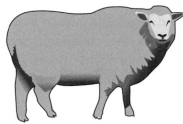

All fibres, both natural and synthetic, contain long-chain molecules called polymers. The diagram opposite shows the long polymer molecules that make up a fibre.

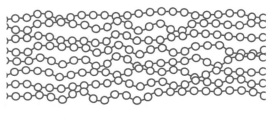

You will learn more about polymers on pages 54 and 55.

Properties of fibres

The properties of a fibre depend on the type of polymer molecules they contain.

Fibres used in making clothes need to be **strong**. This strength comes from the strong bonds inside each polymer molecule and the attractions between the polymer molecules as they line up close together.

Some fibres, like cotton and wool, **absorb water** very readily while others, like nylon and Terylene, do not. This explains why a cotton tee-shirt is more comfortable to wear on hot days than one made of nylon. The cotton absorbs the perspiration and leaves the skin dry and cool. Nylon, on the other hand, is not a good absorber of water and leaves the skin hot and 'sweaty'.

The difference in the ability to absorb water also explains why cotton and woollen clothes take longer to dry than clothes made from nylon or Terylene.

Synthetic fibres can be used to make fabrics with specific properties. For example, fibres of a polymer called Kevlar are so strong that they are used in making bullet-proof vests.

Treating fabrics

Fabrics can be treated in various ways in order to improve their appearance or to improve their properties. Some of these ways are described below.

Dyeing

Fibres tend to be white or dull in colour but they can be dyed in order to make brightly coloured clothes. For a chemical to be suitable as a **dye**, it must be coloured and it must stick to the fibres so that the fibres take on the colour of the dye. Natural dyes from plants can be used but they are not **fast**. This means they don't stick permanently to the fibres and as a result, clothes quickly fade on repeated washing. Natural dyes also fade when exposed to sunlight.

Most dyes that are used nowadays are synthetic and, unlike natural dyes, they are fast: they stick permanently to the fibres and don't wash out, and they keep their colour when exposed to sunlight.

a dye called indigo was used to give these jeans their blue colour.

Flame-proofing

Many people have been badly burned and even killed when their clothes were set alight by standing too close to open fires. All fabrics burn, but cotton is the worst. It catches fire within about five seconds of being exposed to a flame. Fabrics, however, can be treated with chemicals which make them **flame-proof** and this prevents them from burning.

The clothes worn by a fire fighter are both flame-proof and water-proof.

Water-proofing

Outdoor clothes such as cagoules are treated with chemicals which are resistant to water. As a result, rain water is not absorbed into the fabric and just rolls off.

Stain-proofing

Fabrics can also be treated with chemicals which make them resistant to stains like oil and grease. The stain is not absorbed into the fabric and can be simply wiped away.

Top Tip Remember that fabrics can be treated to improve their appearance and to make them flame-proof, water-proof or stain-proof.

Quick Test

1. Fibres can be divided into two groups. One group contains natural fibres. What term is used to describe the fibres in the other group, and how are they made?
2. Name three natural fibres.
3. Name the long-chain molecules that are contained in fibres.
4. Dyeing, water-proofing, stain-proofing and flame-proofing are ways of treating fabrics. Present this information in a table with the headings 'To improve appearance' and 'To improve properties'.

To improve appearance	To improve properties
dyeing	water-proofing
	stain-proofing
	flame-proofing

Answers 1. Synthetic: made by the chemical industry. **2.** Silk, wool and cotton. **3.** Polymers. **4.**

What is a fuel?

In the coal fire shown below, coal is being burned to generate heat energy in order to keep the room warm.

Coal is just one example of what we call a **fuel**. Other examples include wood, petrol, diesel, natural gas and oil.

A **fuel** is defined as any **chemical** which **burns** to produce **energy**. When it burns, the fuel reacts with oxygen from the air. Burning is therefore a chemical reaction and when it takes place, heat energy is released.

Another name for burning is **combustion**.

Top Tip
Make sure you know that a fuel is a chemical which burns to produce energy.

The fire triangle

The coal fire above is being put to good use but many fires, like the forest fire and house fire shown below, can have devastating consequences.

Any **fire** needs three things to be present before it can happen and they are:

- A **fuel** – anything that burns
- **Oxygen** – usually from the air
- **Heat** – a high temperature to start the fire and keep it going.

These three things are shown in **the fire triangle** opposite and if any one of them is taken away, the fire will go out.

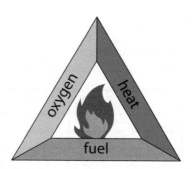

Fire-fighting methods

There are many ways of putting out or extinguishing fires, but each one depends on the removal of one of the sides of the fire triangle, i.e. taking away the fuel or the oxygen or the heat from the fire.

Some examples of fire-fighting methods are given below:

- **Water** can be poured onto the fire. This removes the heat and so puts out the fire. Water is commonly used to put out forest fires and house fires. Water, however, must **not** be used to put out oil, petrol or electrical fires. Oil and petrol float on water and they continue to burn. As a result, the fire spreads rather than being extinguished. Pouring water on a chip pan fire, for example, not only spreads the fire but could cause severe burns when the hot fat or oil spurts out violently from the pan. Using water on an electrical fire could lead to the fire-fighter being electrocuted, since water conducts electricity.

- **A fire blanket** can be thrown over the fire. This starves the fire of oxygen.

- **Sand** can be thrown onto the fire. The covering of sand stops the oxygen reaching the fire.

- **Carbon dioxide** from an extinguisher can be directed at the fire. Carbon dioxide is heavier than air and forms an invisible blanket over the fire. As a result, the fire is deprived of oxygen.

- **Foam** from an extinguisher can be sprayed onto the fire. It prevents oxygen reaching the fire.

Top Tip
You must be able to explain how each fire-fighting method puts out a fire.

Fire-fighting methods employed in the lab and in the home include the use of water, a fire blanket, sand, carbon dioxide and foam.

Quick Test

1. What is a fuel?

2. Give another name for burning.

3. Name the three things that a fire needs.

4. How does covering a chip pan fire with a damp towel put out the fire?

5. Fire breaks are strips of land in a forest from which trees have been removed. How does a fire break prevent a forest fire spreading?

Answers 1. A chemical which is burned to produce energy. **2.** Combustion. **3.** A fuel, oxygen and heat. **4.** Removes the oxygen. **5.** Removes the fuel.

Finite resources and renewable resources

Fossil fuels

Most of Britain's energy comes from burning **fossil fuels**. These are **coal**, **oil**, **natural gas** and **peat**. They are called fossil fuels because they were formed millions of years ago from the remains of dead animals and plants. Coal and peat were formed from plants in swampy areas of the land while oil and natural gas were formed from tiny creatures and plants in the sea.

An oil platform in the North Sea where oil is extracted from underneath the sea bed

Top Tip
The over-use of fossil fuels may lead to a fuel crisis.

Problems with fossil fuels

The main problem with fossil fuels is that they are non-renewable – once formed, they cannot be formed again. They are described as **finite resources** because one day they will run out and can never be renewed. It is estimated that natural gas will last until 2040, oil until 2050 and coal, the most plentiful of the fossil fuels, should last for another 300 years. If we continue to use up fossil fuels at such an alarming rate, then we will be hit by a fuel crisis sooner rather than later.

Another major problem associated with fossil fuels is the pollution caused by spillages from oil tankers and oil refineries. These spillages cause great damage to the environment and to marine life. The oil spoils beaches and kills plants, fish and birds.

Hydrocarbons

The main compounds present in oil and natural gas are known as hydrocarbons. A **hydrocarbon** is a compound which contains hydrogen and carbon **only**. For example, methane (CH_4) is a hydrocarbon but ethanol (C_2H_6O) is not. We can see from its formula that ethanol contains hydrogen and carbon but it cannot be classified as a hydrocarbon because these are not the **only** elements present – it contains oxygen as well.

Hydrocarbons burn in a plentiful supply of air to form water and carbon dioxide. This is known as complete combustion. There is enough oxygen present in the air for all of the hydrogen in the hydrocarbon to be converted into water (H_2O) and all of the carbon to be converted into carbon dioxide (CO_2). The following word equation shows the complete combustion of methane, the main constituent of natural gas.

methane + oxygen → carbon dioxide + water

The apparatus shown opposite can be used to demonstrate that a hydrocarbon burns in a plentiful supply of air to form water and carbon dioxide.

The liquid that collects in the first test tube freezes at 0°C and boils at 100°C and so must be water. The gas that bubbles through the second test tube turns the lime water milky and so must be carbon dioxide since this is the test for carbon dioxide.

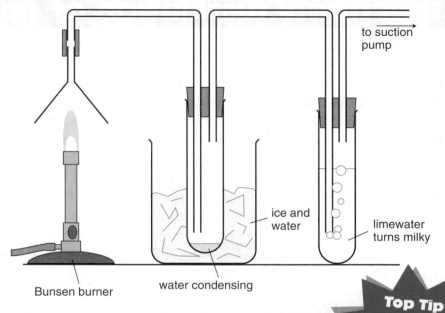

to suction pump

ice and water

limewater turns milky

Bunsen burner

water condensing

Top Tip
Remember that a hydrocarbon contains only hydrogen and carbon.

Renewable sources of energy

Fossils fuels are not renewable and will eventually run out. This has led to the development of fuels which are renewable and will not run out. Some examples of renewable fuels are shown below:

- **Methane** is a renewable fuel which is present in **biogas**. Biogas is formed when bacteria break down plant waste or animal manure in the absence of oxygen. Biogas is an important source of energy in many developing countries.

- **Ethanol** can be made from sugar cane by a process called fermentation. You will find out more about fermentation on page 81. The ethanol produced can be mixed with petrol and used as a fuel for cars.

- **Hydrogen** can be obtained from water by passing electricity through it. At the moment, hydrogen has limited use as a fuel for cars, but in the future, it is likely to be one of the more important renewable fuels.

Top Tip
Remember that methane, ethanol and hydrogen are examples of renewable sources of energy.

Quick Test

1. Name four fossil fuels.

2. Fossil fuels are finite. What does this mean?

3. What is a hydrocarbon?

4. Name the products formed when a hydrocarbon burns in a plentiful supply of air.

5. Methane is found in both natural gas and biogas. Which of these sources of methane is renewable?

Answers 1. Coal, oil, natural gas and peat. **2.** They will run out or they cannot be renewed. **3.** A compound which contains only hydrogen and carbon. **4.** Carbon dioxide and water. **5.** Biogas.

Important processes

Fractional distillation of crude oil

Crude oil is a mixture of hundreds of different hydrocarbons. Unlike the other fossil fuels, crude oil cannot be used directly as a fuel. It has to be separated into **fractions** before it is of any use. This is done by a process called **fractional distillation**. Different hydrocarbons have different boiling points and it is this that allows the fractions to be separated. Each **fraction** still contains a mixture of hydrocarbons, but there are fewer of them. All of the hydrocarbons in a fraction have boiling points within a certain temperature range. The fractional distillation of crude oil is carried out in a fractionating tower which is illustrated below. Alongside the tower you can see the name of each fraction and its uses.

Top Tip
Differences in the boiling points of the hydrocarbons in crude oil allow the fractions to be separated.

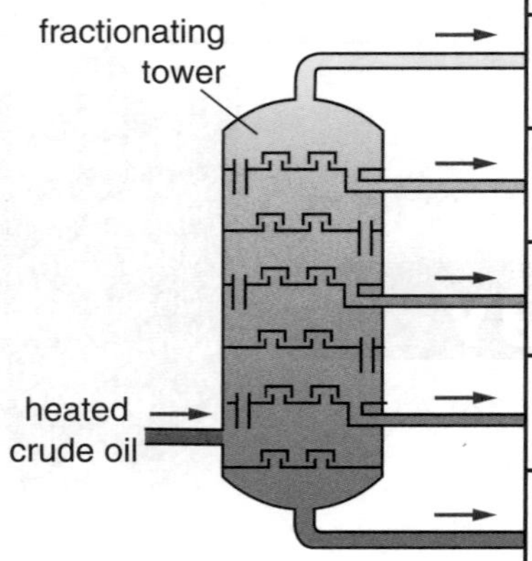

Fraction	Uses
Refinery gas	Fuels for cooking and heating
Naphtha	To make petrol and other chemicals
Kerosene	Aviation and heating fuels
Gas oil	To make diesel
Residue	To make fuel oil for ships and power stations; lubricating oils and waxes; tar for roads and roofing

fractionating tower

heated crude oil

Properties of the hydrocarbons

The table below summarises the trends in some of the **properties** of the fractions separated from crude oil.

Fraction	Size of molecules	Boiling point	Viscosity	Ease of evaporation	Flammability
Refinery gas	Small	Low	Low	Easy	High
Naphtha					
Kerosene					
Gas oil					
Residue	Large	High	High	Difficult	Low

As we move from the refinery gas fraction down to the residue fraction, the size of the hydrocarbon molecules increases. These differences in molecular size can be used to explain the differences in the properties of the fractions.

The **boiling point** and **viscosity** (the thickness of a liquid) increase as the size of the molecules increase. This is because the bonds between the molecules get stronger as the molecules get bigger.

The **ease of evaporation** and **flammability** also depend on the size of the molecules. The smaller the molecules, the easier it is for the fraction to change from a liquid into a vapour, i.e. evaporate. Similarly, the smaller the molecules, the more flammable the fraction, i.e. the more easily it catches fire.

The uses of fractions are related to their properties. For example, petrol, which is made from the naphtha fraction, has a low viscosity so that it can flow readily through the pipes from the tank to the car engine. It is also flammable, so it catches fire easily in the engine.

Cracking

Petrol is one of the more important fuels and the demand for it is huge. It is made from the naphtha fraction. The fractional distillation of crude oil, however, does not produce enough naphtha to satisfy demand. Fractional distillation of crude oil yields more of the larger, long-chain hydrocarbons than are useful for present-day purposes. This is why chemists have devised a process of breaking down these large hydrocarbons into smaller ones. Some of the smaller hydrocarbons are suitable for making petrol. This process is called **cracking**. So, cracking is a chemical reaction in which large hydrocarbons are broken down into a mixture of smaller, more useful hydrocarbons.

Top Tip
In the cracking process, large hydrocarbon molecules are broken down into smaller, more useful molecules.

Quick Test

1. Why can the hydrocarbons in crude oil be separated into fractions?

2. What is a fraction?

3. Why do the hydrocarbons in the naphtha fraction have a lower boiling point than those in the gas oil fraction?

4. Which of the following statements is true?
   A   The residue fraction is less viscous than the naphtha fraction.
   B   The kerosene fraction is more flammable than the residue fraction.
   C   The gas oil fraction evaporates more readily than the naphtha fraction.

5. What is meant by cracking?

Answers 1. Because they have different boiling points. **2.** A group of hydrocarbons with boiling points within a certain temperature range. **3.** Because the molecules in naphtha are smaller in size. **4. B. 5.** Cracking is a process in which large hydrocarbon molecules are broken down into smaller, more useful hydrocarbons.

Pollution problems

What is a pollutant?

A **pollutant** is any substance which damages the environment and harms living organisms within the environment. The land, the water and the air can all be affected by pollution but in this section, we will focus on air pollution.

Air pollution

Listed below are some of the major pollutants of the air.

- **Carbon monoxide and soot (carbon) particles**
  When hydrocarbons burn in a plentiful supply of air, carbon dioxide and water are formed. However, when they burn in a limited supply of air, they undergo **incomplete combustion**. The hydrogen in the hydrocarbon is still converted into water but there is not enough oxygen to convert the carbon into carbon dioxide. Instead, carbon (soot) and carbon monoxide are produced. **Carbon monoxide** and **soot (carbon) particles** are formed when the fuels, petrol and diesel, burn in the cylinders of engines. The reason for this is that the supply of air is limited in engine cylinders. Carbon monoxide is extremely dangerous because it is a very poisonous gas. When soot (carbon) particles are breathed in, the lungs can be badly damaged.

- **Nitrogen dioxide**
  Nitrogen dioxide is another common air pollutant. It is a poisonous gas and irritates the lungs when breathed in. It also dissolves in rain water to form acid rain. Nitrogen dioxide is formed in the cylinders of petrol engines when the petrol/air mixture is sparked. As well as igniting the petrol, the spark provides enough energy to allow some of the nitrogen and oxygen in the air part of the mixture to react and form nitrogen dioxide.

> **Top Tip**
> Carbon monoxide and soot particles are formed when fuels burn in a limited supply of air.

- **Sulphur dioxide**
  Many fuels contain small amounts of sulphur-containing compounds and when they burn, **sulphur dioxide** is formed and released into the air. It is a poisonous gas and irritates the lungs and, like nitrogen dioxide, sulphur dioxide is another cause of acid rain.

- **Lead**
  Lead compounds used to be added to petrol to help it burn more efficiently and when leaded petrol burned, these **lead compounds** were released into the atmosphere. They were major pollutants and caused brain damage to those who breathed in large amounts.

- **Benzene**
  Leaded petrol was slowly phased out and eventually banned in the UK in 2000. The structures of the hydrocarbons in petrol had to be changed so that the petrol would continue to burn efficiently without the lead compounds. One of the new hydrocarbons used was **benzene** but it is toxic (poisonous). People filling their tanks at petrol stations were exposed to these poisonous benzene fumes and this led to benzene being removed from unleaded petrol.

Catalytic converters

One way in which air pollution caused by the burning of petrol can be reduced is by fitting **catalytic converters** to the exhaust systems of cars. In petrol engines, the pollutants carbon monoxide and nitrogen dioxide are produced. When they pass through the catalytic converter, they are converted into the less harmful gases carbon dioxide and nitrogen. The reactions taking place in converters are catalysed by expensive metals like platinum.

Top Tip
Remember that catalytic converters speed up the conversion of pollutant gases into less harmful gases.

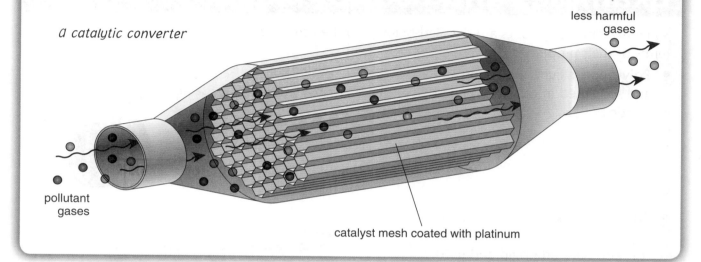

a catalytic converter

less harmful gases

pollutant gases

catalyst mesh coated with platinum

Quick Test

1. Why are carbon monoxide and soot particles formed when diesel burns in an engine?

2. How is nitrogen dioxide produced in a petrol engine?

3. Why was leaded petrol banned?

4. Which two pollutant gases are the main causes of acid rain?

5. What can be used to reduce air pollution from petrol engines?

Answers 1. Because the diesel burns in a limited supply of air or undergoes incomplete combustion. **2.** Nitrogen dioxide is produced when the nitrogen and oxygen from the air react when the petrol/air mixture is sparked. **3.** Because lead compounds are toxic (poisonous) and can cause brain damage. **4.** Nitrogen dioxide and sulphur dioxide. **5.** Catalytic converters.

Uses of plastics

What are plastics?

Plastics are described as **synthetic** materials because they are manufactured by the chemical industry. Most plastics are made from chemicals derived from crude oil.

Examples and uses of plastics

Some examples of plastics are detailed below and the uses to which they are put are explained in terms of their properties.

- **Polythene** is light and strong and so is used in carrier bags and plastic bottles.

- **Polystyrene** is used in hot-drink cups because it is a heat insulator. It is also hard and rigid and used for radio and TV casings.

- **PVC** is used in guttering and window frames because it is strong and rigid. It is an electrical insulator and so is used in a more flexible form as a covering for electrical wires.

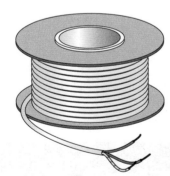

- **Perspex** is used in motorcycle windshields and solar panels because it is transparent, hard and shatterproof.

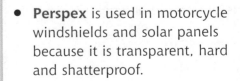

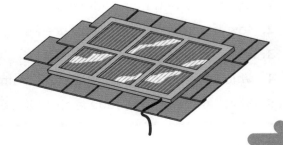

- **Nylon** is strong, hard-wearing and flexible and used in toothbrushes and in clothes.

- **Kevlar** is used in bullet-proof vests because it is very strong and in fire-fighters' clothing since it is heat resistant.

- **Silicones** repel water and this explains why they are used as sealants in baths and showers.

Bakelite and **formica** are another two common plastics. Bakelite is used in electrical plugs and sockets and inside electrical fittings because it is an electrical insulator and is heat resistant. Formica too is heat resistant and it is also very hard. This is why it is used in kitchen work-tops.

Top Tip
You must be able to link the everyday uses of plastics to their properties.

Quick Test

1. Most plastics are made from which raw material?

2. Which line in the table shows suitable properties of a plastic which could be used in greenhouses instead of glass?

3. Why can PVC be used as covering for electrical wires?

4. Give one important property of
   a) silicones   **b)** Kevlar.

	Lets light through?	Effect of heat	Effect of light
A	yes	none	very little
B	no	none	none
C	yes	none	becomes brittle
D	yes	cracks	very little

Answers 1. Crude oil. **2.** A. **3.** Because it is an electrical insulator. **4. a)** repels water; **b)** is very strong.

Advantages and disadvantages

Plastics versus natural materials

For some uses, plastics have advantages over natural materials and vice versa. For example:

- Window frames made from PVC rather than wood last longer and don't need to be painted. Wood, however, has the advantage that it is a renewable resource.
- A bucket made of polythene rather than steel is lighter, less expensive and does not rust. Steel, however, has the advantage that it is stronger and is less affected by heat.

Plastics and pollution

Pollution from litter

Natural materials, such as wood, paper and food waste, are **biodegradable**. This means that they are broken down by bacteria in the soil and eventually rot away. Most plastics, however, are not biodegradable and will remain unchanged for many years. They will therefore pollute the environment if thoughtlessly thrown away. As well as being unsightly, plastic litter can cause death and a great deal of suffering to animals.

Top Tip
Biodegradable materials are broken down by bacteria in the soil and rot away.

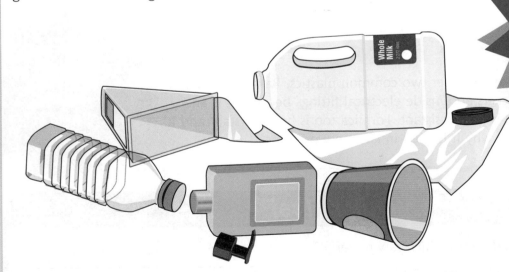

To reduce the litter problem, chemists have developed some biodegradable plastics. One such plastic is **Biopol**.

Pollution from burning

When plastics burn or smoulder they give off toxic (poisonous) fumes. All plastics contain carbon and when they burn in a limited supply of air, the carbon burns to form the poisonous gas, carbon monoxide.

Disposal of plastic waste

Ways of disposing of plastic waste include:

- **Incineration (burning)**

  Plastic waste can be burned or incinerated and in some cases, the energy released in the process is used for heating. However, a major problem with burning plastic waste is the production of poisonous fumes, like carbon monoxide, which pollute the air.

- **Recycling**

  Nowadays, more and more plastic waste is being recycled. This involves reprocessing the plastic and making it into useful products. For example, waste plastic from farms can be recycled into garden furniture.

  garden furniture made from plastic waste

  There are many different plastics and for recycling to be successful, the waste has to be sorted out into the different types of plastic. This can be difficult and expensive and explains why a relatively small proportion of plastic waste is recycled.

  Plastics are made from chemicals derived from crude oil and recycling them is important because it helps to conserve this finite resource. Chemists are also looking for renewable sources of plastics in order to save oil.

- **Burying**

  A lot of plastic waste is buried in holes in the ground in so-called landfill sites. Plastic waste which is buried is not put to any use. All that happens is that more and more land is being used up in order to accommodate all of the plastic waste.

Top Tip

Remember that options for the disposal of plastic waste include incineration, recycling and burying.

Quick Test

1. Iron pipes are being replaced by polythene pipes to carry mains water. State one major advantage that polythene pipes have over iron pipes.

2. Some biodegradable plastics have been developed. What is meant by a **biodegradable** plastic?

3. Name one of the gases present in the toxic fumes that are given off when plastics burn or smoulder.

4. Give three ways of disposing of plastic waste.

5. Which method of disposing of plastic waste can produce poisonous gases?

Answers 1. Polythene pipes do not rust. **2.** A biodegradable plastic is one which is broken down by bacteria in the soil and rots away. **3.** Carbon monoxide. **4.** Incineration (burning), recycling and burying. **5.** Incineration or burning.

Thermoplastic/ thermosetting plastics and making plastics

Thermoplastic and thermosetting plastics

Plastics can be classified as either **thermoplastic** or **thermosetting** according to how they are affected by heat. Those that are thermoplastic soften on heating and can be **reshaped**. Examples include polythene, polystyrene, PVC and nylon. Thermosetting plastics do not soften on heating and so cannot be reshaped. Bakelite and formica are examples of thermosetting plastics. Bakelite is used in electrical **plugs** and sockets because it is an electrical insulator. Formica is used in kitchen work-tops because it is a heat insulator and will not melt when hot pans are placed on it.

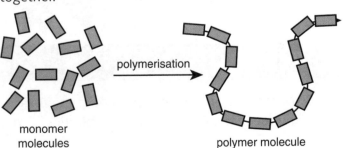

Top Tip
Make sure you know the difference between thermoplastic and thermosetting plastics.

Polymers

Plastics are made up of large long-chain molecules called **polymers**. A polymer molecule is formed when hundreds of small molecules called **monomers** join together.

polymerisation

monomer molecules

polymer molecule

Top Tip
Polymerisation is the process whereby lots of small molecules called monomers join together to form larger polymer molecules.

This process of converting the small monomer molecules into a large polymer molecule is known as **polymerisation**.

Naming polymers

Some polymers can be named by simply writing 'poly' in front of the name of the monomer. For example, the polymer made from the monomer, vinyl chloride, is called poly(vinyl chloride). Some other examples are shown in the following table.

Name of monomer	Name of polymer
chloroethene	poly(chloroethene)
phenylethene	poly(phenylethene)
ethene	poly(ethene)

The polymer **poly(ethene)** is more commonly known as **polythene**.

Just as we can name a polymer given the name of its monomer, we can also name a monomer given the name of the polymer. For example, the polymer called poly(tetrafluoroethene) would be made from the monomer called tetrafluoroethene.

Top Tip

Make sure you know how to name a polymer given the name of the monomer and vice versa.

Quick Test

1. What type of plastic does not soften on heating and cannot be reshaped?

2.

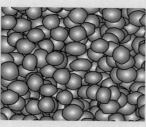

polymer granules

melted and moulded →

plastic bucket

What type of polymer has been used in the above process?

3. What is meant by the term **polymerisation**?

4. Name the polymer made from the monomer styrene.

5. Perspex is the trade name for poly(methylmethacrylate). Name the monomer used to make perspex.

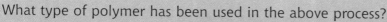

Answers 1. Thermosetting. **2.** Thermoplastic. **3.** Polymerisation is the process that takes place when lots of small monomer molecules join together to form large polymer molecules. **4.** Polystyrene. **5.** Methylmethacrylate.

PPA 1: Electrical conductivity

Introduction

Some substances are **conductors** of electricity. This means that they allow a current of electricity to pass through them. Other substances do not let a current of electricity pass through them and they are called **non-conductors** or electrical insulators. In this PPA, the electrical conductivities of elements, both metals and non-metals, were tested.

Aim

To test the conductivity of some metals and non-metals and from the results work out a general rule about the electrical conductivity of elements.

Procedure

- The following circuit was set up to test the electrical conductivity of the elements aluminium, carbon (graphite), copper, iron, nickel, sulphur and zinc.

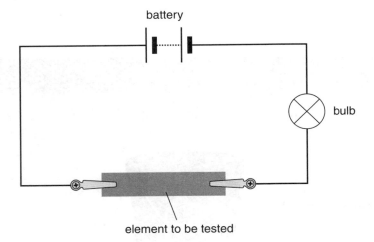

- Each element was identified as a metal or a non-metal and then placed in the circuit to test its electrical conductivity. If the bulb lights up, then the element is an electrical conductor.

Results

As well as having the results of the experiment, the table opposite also includes the names of other elements and their electrical conductivities. These are shown in red in the table. These were not tested in the PPA because of safety reasons.

Element	Metal/non-metal	Conductor/non-conductor
Aluminium	metal	conductor
Carbon (graphite)	non-metal	conductor
Copper	metal	conductor
Iron	metal	conductor
Nickel	metal	conductor
Sulphur	non-metal	non-conductor
Zinc	metal	conductor
Iodine	non-metal	non-conductor
Phosphorus	non-metal	non-conductor
Mercury	metal	conductor
Bromine	non-metal	non-conductor
Selenium	non-metal	non-conductor

Conclusion

Metal elements conduct electricity but non-metal elements do not conduct electricity. The element which does not fit this rule is carbon (graphite) – it is a non-metal but conducts electricity.

Points to note

- A bulb was used in the circuit to test if the element conducted electricity. If the bulb lights up, then the element conducts electricity.
- A buzzer or an ammeter could be used in place of the bulb to find out if the element conducted electricity.
- The Periodic Table on page 8 of the Data Booklet can be used to determine whether an element is a metal or a non-metal. The elements below the dark line are metals.
- Care had to be taken when working with the sulphur since it is flammable. Ignition sources like lighted Bunsen burners must be absent to prevent the risk of the sulphur catching fire.

Quick Test

1. When an element was placed in the circuit, how could you tell that it conducted electricity?

2. What could be used instead of a bulb to show that an element conducts electricity?

3. What is the general rule concerning the electrical conductivities of metals and non-metals?

4. Name a non-metal element which is an electrical conductor.

5. Complete the following table:

Element	Metal or non-metal?	Conductor or non-conductor?
Lithium		
Arsenic		
Silver		
Lead		

Answers 1. The bulb lit up. **2.** A buzzer or an ammeter. **3.** Metals conduct electricity but non-metals do not. **4.** Carbon in the form of graphite. **5.** Lithium, metal, conductor; Arsenic, non-metal, non-conductor; Silver, metal, conductor; Lead, metal, conductor

PPA 2: Reactions of metals with acid

Introduction

Different metals react with acid at different speeds. A metal which reacts quickly with an acid is said to be a **reactive metal** while a metal which reacts only slowly or does not react at all is an **unreactive metal**. When a metal does react with an acid, bubbles of hydrogen gas are formed and the speed at which the bubbles are given off gives a measure of how reactive the metal is.

Aim

To place the metals zinc, magnesium and copper in order of reactivity by observing how fast they react with dilute hydrochloric acid.

Procedure

- Some dilute hydrochloric acid was added to each of three test tubes to a depth of about 4 cm.
- A piece of zinc was added to the first test tube, a piece of magnesium to the second test tube and a piece of copper to the third.
- In each case, it was noted whether bubbles of gas were produced or not and if they were, the speed at which they were given off.

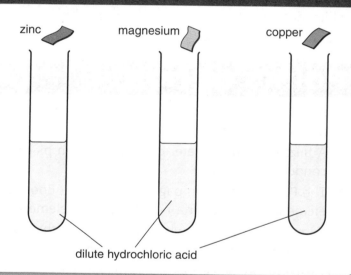

dilute hydrochloric acid

Results

Metal	Bubbles of gas produced?	Reaction speed
Zinc	yes	slow
Magnesium	yes	fast
Copper	no	no reaction

Conclusion

The order of reactivity of the metals, starting with the most reactive, is magnesium, zinc and copper.

Points to note

- To make the experiment fair, the only factor that was changed was the metals used. Other factors were kept the same. For example:

  (a) the acid, i.e. hydrochloric acid

  (b) the concentration of the hydrochloric acid

  (c) the volume of the acid

  (d) the temperature of the acid

  (e) the size of the piece of metal foil used

- When a metal reacts with acid, hydrogen gas is produced. We got some idea of the reactivity of each of the metals by the speed at which the bubbles of hydrogen gas were given off.

- The magnesium and the hydrogen gas produced are flammable. This means that all ignition sources such as lighted Bunsen burners must be absent to prevent the risk of fire.

- Care must be taken not to breathe in the acid mist that is formed when magnesium reacts with dilute hydrochloric acid. The acid mist irritates the nose, throat and lungs.

Quick Test

1. Name the gas produced when metals react with dilute hydrochloric acid.

2. How can you get an idea of the reactivity of a metal when it is added to dilute hydrochloric acid?

3. Apart from copper, name another metal which will **not** react with dilute hydrochloric acid. You may wish to use page 6 of the Data Booklet to help you.

4. Three metals, iron, calcium and lead, were added to dilute sulphuric acid. Using information from page 6 of the Data Booklet, complete the following table.

Metal	Bubbles of gas produced?	Reaction speed
Iron		
Calcium		
Lead		

5. Why was it important not to breathe in the acid mist that was formed when magnesium reacted with dilute hydrochloric acid?

5. Because it irritates the nose, throat and lungs.

Metal	Bubbles of gas produced?	Reaction speed
iron	yes	slow
calcium	yes	very fast
lead	yes	very slow

Answers 1. Hydrogen. **2.** The speed at which the bubbles of hydrogen gas are given off. **3.** Mercury or silver or gold. **4.**

PPA 3: Factors which affect lathering

Introduction

Since substances such as oil and grease do not dissolve in water, it is difficult to clean things in water alone. This is why detergents are used as cleaning chemicals. Detergents break down oil and grease into tiny droplets which mix with the water and can be washed away. Most detergents form a **lather** when they are shaken with water. Two of the factors which might affect the amount of lather formed are:

- the type of detergent used
- the volume of detergent used

Aim (using different types of detergent)

To find out if different types of detergent affect the amount of lather formed when they are shaken with water.

Procedure

- Using a syringe, 3 cm³ of water were added to a test tube followed by 2 drops of automatic washing powder solution. The test tube was then stoppered.
- With a thumb on the stopper, the test tube was shaken hard for 15 seconds.
- The solution was then allowed to settle for 15 seconds and the height of the lather was measured using a ruler.
- The above procedure was repeated again with the same automatic washing powder solution in order to give a duplicate result.
- The whole experiment was repeated with a solution of the non-automatic washing powder and then with a solution of the dishwasher powder.

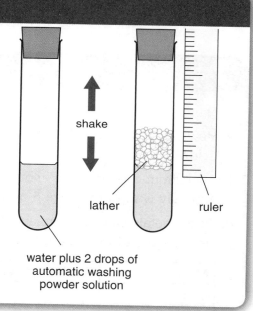

shake

lather   ruler

water plus 2 drops of automatic washing powder solution

Results

Type of detergent solution	Height of lather in cm		
	Result 1	Result 2	Average
automatic washing powder	2·1	2·5	2·3
non-automatic washing powder	5·8	6·2	6·0
dishwasher powder	1·3	1·5	1·4

Conclusion

As the type of detergent changes the amount of lather formed also changes.

Aim (using different volumes of detergent)

To find out if different volumes of detergent affect the amount of lather formed when shaken with water.

Procedure

- Using a syringe, 3 cm^3 of water were added to a test tube followed by 1 drop of washing-up liquid solution. The test tube was then stoppered.
- With a thumb on the stopper, the test tube was shaken hard for 15 seconds.
- The solution was then allowed to settle for 15 seconds and the height of the lather was measured using a ruler.
- The procedure was repeated again with 1 drop of washing-up liquid solution.
- The whole experiment was repeated with 2 drops of washing-up liquid solution and then with 3 drops.

Results

Number of drops of washing-up liquid solution	Height of lather in cm		
	Result 1	Result 2	Average
1	5·9	6·1	6·0
2	9·3	9·3	9·3
3	12·5	12·9	12·7

Conclusion

As the number of drops of washing-up liquid solution increased, the amount of lather formed also increased.

Points to note

- To make each investigation fair, only one factor was changed. In the **first investigation**, the type of detergent solution was changed but the volume of the detergent solution was kept the same. In the **second investigation**, the volume of detergent solution was changed but the type of detergent was kept the same.
- In both investigations, the temperature of the water, the volume of the water and the 'shaking' time were kept the same. The same test tube or identical test tubes were also used.
- To get some idea of the amount of lather formed, the height of the lather was measured using a ruler.
- Each experiment was carried out in duplicate. This was done to make sure that the results were valid and reliable.
- Care had to be taken when working with the detergent solutions since they irritate the eyes.

Quick Test

1. When investigating the effect of volume of detergent on the amount of lather formed, state **five** factors which had to be kept the same.

2. How can you get an idea of the amount of lather formed when a detergent is shaken with water in a test tube?

3. Why was each experiment carried out in duplicate?

4. Why is it necessary to wear safety glasses when working with detergent solutions?

Photosynthesis, respiration and the greenhouse effect

Photosynthesis

Plants make their own food by a process called **photosynthesis**. Photosynthesis takes place in the leaves of green plants and involves the reaction between **carbon dioxide** and **water** to produce **glucose** and **oxygen**:

carbon dioxide + water → glucose + oxygen

Carbon dioxide from the air is absorbed through the leaves of the plants and water is drawn up through the roots. The oxygen produced in the process is released through the leaves into the atmosphere. Some of the glucose formed is used by the plant as a source of energy and the rest is converted into starch and cellulose.

Energy is needed for photosynthesis to take place and this comes from the Sun. A green-coloured chemical called **chlorophyll** is also essential for the process. It is present in the green parts of plants and its role is to absorb the light energy from the Sun.

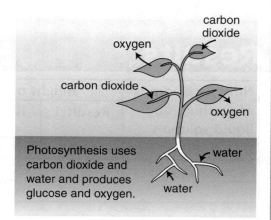

Photosynthesis uses carbon dioxide and water and produces glucose and oxygen.

Respiration

Respiration is the process whereby all living things, both plants and animals, get energy from food. It is a chemical reaction in which **glucose** reacts with **oxygen** from the air and the **carbon dioxide** and **water** that are formed are released back into the atmosphere. Animals obtain the glucose needed for respiration by eating food which has come from plants.

The word equation for respiration is shown below:

glucose + oxygen → carbon dioxide + water

The energy released in the process of respiration is used by animals in a number of ways e.g. to keep warm and to move around.

By looking at the word equations for respiration and photosynthesis, you'll notice that each is the reverse of the other. It is important to remember that respiration can take place in both animals and plants, but photosynthesis can only take place in plants.

Photosynthesis uses up carbon dioxide and releases oxygen into the air, while respiration uses up oxygen and releases carbon dioxide into the air. Clearly, photosynthesis and respiration are important in maintaining the balance between oxygen and carbon dioxide in the air.

Top Tip
Respiration is the reverse of photosynthesis. Energy is given out during respiration and taken in during photosynthesis.

The greenhouse effect

The carbon dioxide in the air plays an important role in keeping the Earth warm. This is known as the **greenhouse effect** because the carbon dioxide in the air absorbs some of the Sun's energy just as the glass in a greenhouse traps the Sun's energy and heats up the air inside it. If the amount of carbon dioxide in the atmosphere remains constant, the temperature of the atmosphere should remain constant. However, because of man's interference, the delicate balance between carbon dioxide and oxygen in the atmosphere has been disturbed. With the clearing of rain forests, less carbon dioxide is removed from the atmosphere by photosynthesis and with the increased combustion of fossil fuels (coal, oil and natural gas) more carbon dioxide enters the atmosphere. As a result, the proportion of carbon dioxide in the atmosphere has increased and this build-up causes the atmosphere to trap more of the Sun's energy and raise its temperature. This is known as **global warming**.

Clearing rain forests in South America.

Top Tip
An increase in the level of carbon dioxide in the atmosphere is believed to be the cause of global warming.

Quick Test

1. Which two substances react together in the process of photosynthesis?
2. Name the chemical in plants which traps the Sun's energy.
3. Write the word equation for respiration.
4. What is the relationship between photosynthesis and respiration?
5. Give two reasons why the level of carbon dioxide in the atmosphere has increased.

Answers 1. Carbon dioxide and water. **2.** Chlorophyll. **3.** glucose + oxygen → carbon dioxide + water. **4.** One is the reverse of the other. **5.** Clearing rain forests and the increased combustion of fossil fuels.

Using chemicals to save plants

Feed the world

The rapid increase in the population of the world has meant an ever-increasing demand for food. One of the more important food types that people need are proteins. We get protein by eating plants (vegetables, fruit etc.) and by eating animals (meat, fish, etc.). To meet the growing demand for food, there is a need to grow more plant crops and increase their yield.

Treating plant crops

One way of improving the growth and yield of plant crops is to treat them with artificial chemicals made by the chemical industry. The three main types of artificial chemicals used to keep plant crops healthy are:

- **Pesticides**
  Plant crops can be eaten by **pests** such as insects, slugs, snails, caterpillars and worms.

  Pesticides can be sprayed on the plants in order to kill and control these pests.

- **Fungicides**
  Diseases caused by **bacteria** and **fungi** can also attack plants. These diseases rapidly spread through crops and destroy them. Examples of fungal diseases include 'blight' in potatoes and 'rust' in wheat.

  'Rust' disease is so-called because it has a brown colour just like rust on iron.

  Fungal diseases can be prevented by spraying the crops with **fungicides**.

- **Herbicides**
  Weeds are another problem faced by farmers. Weeds take valuable nutrients from the soil which means the crop will be deprived of them. These weeds can be killed, however, by spraying the crops with **herbicides**.

Top Tip
Remember that
- pesticides control pests
- fungicides prevent diseases
- herbicides kill weeds.

Problems with pesticides

Pesticides have been used for many years. They were considered to be a safe and effective way of getting rid of pests but many of them proved to be **toxic** (poisonous) to humans and other animals. DDT, used to kill insects, is one such harmful pesticide. It was banned in the UK as long ago as 1984. The pesticides that are still in use today have to be handled with a great deal of care due to their toxic nature.

More and more alternatives to toxic pesticides are being introduced into farming and horticulture. One way of controlling pests is to use natural predators. For example, ladybirds are used commercially to control infestations of greenfly in greenhouses. The ladybirds eat the greenfly and so protect the plants.

Top Tip
Pesticides are toxic (poisonous) and must be used with great care.

Quick Test

1. Why is there a demand for more and more food?
2. What type of chemical is used to kill insects and slugs?
3. What causes plants to become diseased?
4. Name the type of chemical that is used to kill weeds?
5. Why must pesticides be used with care?

Answers 1. Because of the growing population of the world. **2.** A pesticide. **3.** Bacteria and fungi. **4.** A herbicide. **5.** Because they are toxic (poisonous).

Fertilisers

Essential elements

The three most important elements needed for healthy plant growth are **nitrogen** (**N**), **phosphorus** (**P**) and **potassium** (**K**). These essential elements are not supplied to the plant as free elements but as compounds, i.e. combined with other elements. The compounds that they are in must be **soluble** in water so that they will dissolve and be drawn up into plants through their roots.

Top Tip
Remember that nitrogen, phosphorus and potassium are the three essential chemical elements needed for healthy plant growth.

Natural fertilisers

As plants grow, they remove the essential elements from the soil but in areas of natural vegetation, such as woodlands, the plants die and decay during winter and the essential elements are returned to the soil. In addition, any animals which die in these areas also decay and return essential elements to the soil. In cultivated areas, however, plant crops are harvested and so the essential elements are not returned to the soil. As a result, soil fertility decreases but it can be restored by adding **natural fertilisers** such as animal manure and compost.

Artificial fertilisers

As the world population increased, the demand for food grew and crop production intensified. The supply of natural fertilisers proved insufficient to keep the soil fertile and this led to the production of **artificial fertilisers**. Artificial fertilisers are made by the chemical industry. They must be **soluble** in water so that they can be drawn up through the roots of plants, and they must contain one or more of the three essential elements nitrogen, phosphorus and potassium. The major artificial fertilisers are ammonium compounds, potassium compounds, nitrate compounds and phosphate compounds. The following table shows some examples of artificial fertilisers and the essential elements they contain.

Artificial fertiliser	Essential elements present		
	nitrogen	phosphorus	potassium
Ammonium nitrate	✓	✗	✗
Potassium nitrate	✓	✗	✓
Ammonium phosphate	✓	✓	✗

Pollution problems

Although artificial fertilisers are important in improving crop production, a number of environmental problems arise through their use. One such problem caused by nitrate fertilisers is the pollution of rivers and lochs and public water supplies. Since fertilisers are soluble in water, they are easily washed out of the soil, and as a result the nitrate levels in water have increased. These nitrates are harmful to humans if they are present in drinking water. Large quantities of nitrates in rivers and lochs cause an increase in the growth of algae, and algae remove oxygen from the water and leave it lifeless.

Top Tip
Nitrate fertilisers can pollute rivers, lochs and public water supplies.

Nitrate-making plants

The vast majority of plants obtain the essential element nitrogen by absorbing nitrate compounds from the soil. There are, however, a few plants such as **peas**, **beans** and **clover** which can make their own nitrates. They have lumps on their roots known as **root nodules**. Inside these nodules, nitrogen from the air is converted into nitrates. When these plants are ploughed back into the ground, the fertility of the soil is increased.

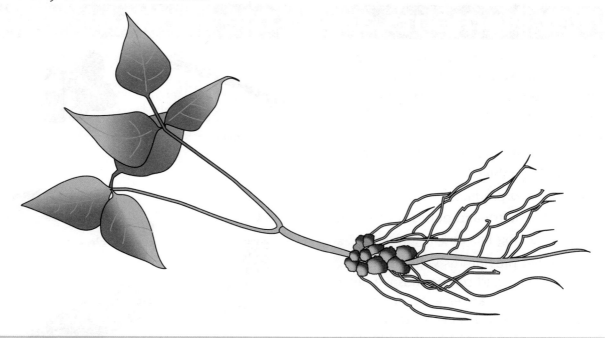

Quick Test

1. Name the three elements which are essential for healthy plant growth.

2. Give two examples of natural fertilisers.

3. Why must fertilisers be soluble in water?

4. What can cause the water in rivers to become lifeless?

5. Name three plants which can convert nitrogen from the air into nitrates.

Answers 1. Nitrogen, phosphorus and potassium. **2.** Animal manure and compost. **3.** Because they must be in solution before they can be taken in through the roots of plants. **4.** The presence of large quantities of nitrates. **5.** Peas, beans and clover.

Elements in the body

What are the main elements in the human body?

Just four elements make up 96% of the human body . These are **oxygen**, **carbon**, **hydrogen** and **nitrogen**, and the following table shows the percentage (%) by mass of each.

Element	Percentage (%) by mass
Oxygen	65
Carbon	18
Hydrogen	10
Nitrogen	3

Top Tip
Make sure you remember that water makes up more than 60% of body weight.

It is important to note that these elements are not present in the body in the uncombined state but are present in compounds, e.g. water, proteins, fats, carbohydrates such as glycogen and glucose, DNA etc. The most abundant compound is water, which makes up more than 60% of body weight.

What elements are in the foods we eat?

To maintain a healthy body it is important that we have a **balanced diet**. This means eating a variety of foods from the five basic food groups but in the correct proportions. These five basic food groups are:

- bread, potatoes, pasta, rice and cereals
- fruit and vegetables
- milk and dairy products
- meat, fish, poultry and pulses such as lentils
- foods containing fats and sugars

A balanced diet provides us with all the essential elements and compounds we need. The compounds that are essential in our diets are **carbohydrates**, **fats** and **proteins**. The main elements present in these essential compounds are shown in the following table.

Essential compound	Elements present			
	oxygen	carbon	hydrogen	nitrogen
Carbohydrates	✓	✓	✓	✗
Fats	✓	✓	✓	✗
Proteins	✓	✓	✓	✓

So, foods which contain carbohydrates, fats and proteins provide us with the four main elements in our bodies i.e. oxygen, carbon, hydrogen and nitrogen.

Minerals and trace elements

As well as carbohydrates, fats and proteins, our bodies need small amounts of **minerals** to supply us with other important elements. For example, **calcium**, which is present in milk, is needed for healthy bones and teeth; **iron**, which is present in red meats and some vegetables and cereals, is needed for healthy blood.

Milk supplies calcium and meat supplies iron.

Minerals also supply us with **trace elements**, so-called because we only need tiny amounts of them. Two examples of these trace elements are zinc and iodine – zinc is essential for growth and iodine is needed in the thyroid gland. Some trace elements can be toxic (poisonous) if taken in too large quantities. For example, too much copper in the diet can cause damage to the liver.

Top Tip
Remember that minerals supply the body with calcium for bones and teeth and iron for the blood as well as the trace elements that the body needs.

Quick Test

1. Name the four main elements in the human body.

2. Are elements present in the body as compounds or as the free elements?

3. What is the most abundant compound in the body?

4. Name three essential compounds for a healthy diet.

5. Which element is present in proteins but not in carbohydrates or fats?

6. Minerals in our diet supply us with small quantities of important elements. Name the element needed for
   a) bones and teeth
   b) blood.

Answers 1. Oxygen, carbon, hydrogen and nitrogen. **2.** As compounds. **3.** Water. **4.** Carbohydrates, fats and proteins. **5.** Nitrogen. **6. a)** Calcium: **b)** Iron.

Different carbohydrates

What are carbohydrates?

One group of essential compounds in the food we eat are made up of **carbohydrates**. Carbohydrates are made by plants and they are an important part of our diet because they are used by the body to produce energy. Examples of foods which are rich in carbohydrates include cereals, bread, potatoes, pasta, milk, chocolate, biscuits, cakes, honey and jams.

Carbohydrates are compounds which contain the elements carbon, hydrogen and oxygen.

Classifying carbohydrates

Carbohydrates can be divided into two main groups. These are **sugars** and **starch**. Sugars are sweet and they dissolve in water, forming clear solutions. Starch, on the other hand is not sweet and does not dissolve readily in water – it forms a slightly cloudy solution. Sugars are made up of small molecules, while starch contains very large molecules.

Important examples of sugars include **glucose**, **fructose**, **maltose** and **sucrose**. Sucrose is the carbohydrate found in ordinary table sugar.

Top Tip
Carbohydrates can be divided into two main groups – sugars and starch.

Benedict's test

Most sugars when heated with blue **Benedict's solution** turn it cloudy orange.

Of the carbohydrates glucose, fructose, maltose, sucrose and starch, only **glucose**, **fructose** and **maltose** turn Benedict's solution cloudy orange – sucrose and starch do not. This means that glucose, fructose and maltose can be distinguished from sucrose and starch by using Benedict's test.

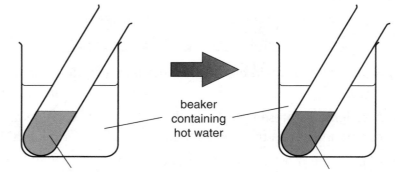

beaker containing hot water

blue Benedict's solution + glucose or fructose or maltose

The sugar has turned the Benedict's solution from blue to cloudy orange.

The iodine test

When brown iodine solution is added to starch, a blue-black colour appears.

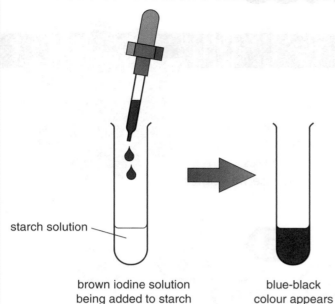

starch solution

brown iodine solution
being added to starch

blue-black
colour appears

Top Tip
Remember that blue Benedict's solution turns cloudy orange when heated with glucose or fructose or maltose and brown iodine solution turns blue-black when added to starch.

Starch is the only compound that turns iodine solution blue-black, so this means that iodine solution can be used to test for starch and to distinguish it from other carbohydrates.

Quick Test

1. How are carbohydrates used by the body?

2. Which elements are present in carbohydrates?

3. State three differences between glucose and starch.

4. How would you distinguish between glucose and sucrose?

5. Name the solution used to test for starch and state the colour change in this solution when it reacts with starch.

Answers 1. To produce energy. **2.** Carbon, hydrogen and oxygen. **3.** Glucose is sweet, soluble in water and is made up of small molecules, while starch is not sweet, not very soluble in water and is made up of large molecules. **4.** Heat each of them with Benedict's solution. Glucose will turn it cloudy orange and sucrose will have no effect. **5.** Iodine solution; brown to blue-black

Reactions of carbohydrates

Making starch

Plants take carbon dioxide from the air and water from the soil and convert them into glucose by the process of photosynthesis. These glucose molecules, like all sugar molecules, are small in size and lots of them join together to form starch molecules, which are very large. This process, which also takes place in plants, is an example of polymerisation. The glucose monomer units are converted into the polymer, **starch**.

glucose monomers

polymerisation

starch polymer

Up to a million glucose molecules are needed to make just one starch molecule.

Plants use the starch they make as **an energy store**.

Breaking down starch in the body

Before we can make use of the starch in foods such as potatoes, bread and pasta, it has to be broken down in our bodies. The starch molecules react with water present in our digestive juices and break down into smaller glucose molecules. This process of digestion is catalysed (speeded up) by **enzymes** and **acid**, which are also present in the digestive juices. Unlike the large starch molecules, the glucose molecules that are formed are small enough to pass through the gut wall into the blood stream. Once there, they are carried to various cells in the body where they undergo respiration to produce the energy we need.

Breaking down starch in the lab

The breakdown of starch into smaller sugar molecules can be demonstrated in the lab using an enzyme and using a dilute acid.

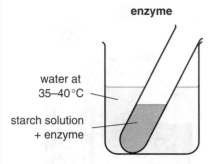

enzyme

water at 35–40°C

starch solution + enzyme

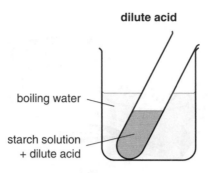

dilute acid

boiling water

starch solution + dilute acid

After the reaction mixtures have been heated for some time, they are tested with Benedict's solution. In both cases, the blue Benedict's solution turned cloudy orange. Since starch does not give a positive test with Benedict's solution, this means that the starch molecules have been broken down into smaller sugar molecules.

The breakdown of starch using an enzyme is much faster than that with dilute acid. This shows that enzymes are very efficient catalysts.

You'll notice in the enzyme-catalysed breakdown of starch, the water bath is at a temperature of 35–40 °C. If boiling water had been used then the starch would not have broken down. The reason for this is that the enzyme would have been destroyed and would no longer have been able to catalyse the reaction. In general, enzymes in the body work best at body temperature, i.e. 37 °C, and at higher temperatures they are destroyed.

Top Tip
Remember that body enzymes function best at body temperature (37°C) and are destroyed at higher temperatures.

Quick Test

1. What monomer units are used to make the polymer starch?

2. Why do plants make starch?

3. During digestion, starch is broken down into which substance?

4. Name two substances which catalyse the breakdown of starch.

5. State the temperature at which body enzymes function best.

Answers 1. Glucose. **2.** To store energy. **3.** Glucose. **4.** Enzymes and dilute acid. **5.** 37°C.

Fats and oils

What are fats and oils?

Fats and oils are another important class of food. Fats are usually solids and are obtained from animals. Examples include lard, dripping and butter.

Oils are liquid at room temperature and are obtained from plants. Examples include sunflower oil, olive oil and corn oil.

You can find information about the composition of different foods in your Data Booklet, page 7. The table gives masses, in grams, of carbohydrate, fat and proteins per 100 grams of food.

The table on the right shows the percentage (%) of fat in some everyday foodstuffs.

The percentage of fat in our foods is usually given on the food labels.

Food	Percentage (%) of fat
Bread	2
Butter	83
Cheese	35
Peanuts	49
Sausages	31

You already know that our bodies use carbohydrates to obtain energy. Fats and oils also provide our bodies with energy. In fact, fats and oils are much more concentrated sources of energy than the same mass of carbohydrate.

This means that 1 gram of a fat or oil will provide much more energy than 1 gram of carbohydrate when eaten.

This can be shown by carrying out the experiment shown in the diagram on the right.

The results of such an experiment are given in the table below.

30 cm³ water

crucible

wick

vegetable oil   sugar lump

Foodstuff	Temperature of water at the start	Temperature of water at the end	Rise in temperature of the water
3 grams of vegetable oil	22 °C	85 °C	63 °C
3 grams of sugar	22 °C	61 °C	39 °C

You can see from the results table that the oil has given out more heat than the sugar.

Top Tip
Fats and oils are more concentrated sources of energy than carbohydrates such as sugars and starches.

Testing for fats and oils

The test for fats and oils is the filter paper test. Both fats and oils leave a greasy, see-through stain on filter paper. This can be shown by putting a drop of vegetable oil on filter paper or wiping butter on to the filter paper. A transparent mark is left on the filter paper. Greasy foods such as pies or foods cooked in oil such as chips also leave the same greasy, see-through stain on filter paper.

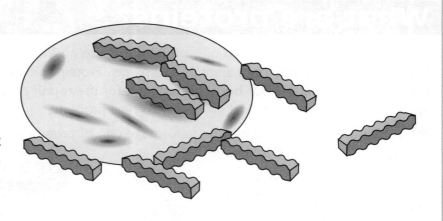

Fats, oils and heart disease

There are different types of fats and oils. Generally speaking oils are liquid and fats are solid at room temperature. Fats and oils are sometimes further divided into **saturates** and **polyunsaturates**.

Saturates are more likely to be present in the more solid fats such as lard and butter and many doctors and scientists believe that they increase the cholesterol level in the blood. This is thought to be responsible for heart disease as doctors find that people who have suffered a heart attack have high levels of cholesterol in their bloodstream. Doctors often test the blood of their patients to find the cholesterol level and may advise them to go on a low cholesterol diet such as avoiding certain fatty foods.

Polyunsaturates are more likely to be present in oils such as sunflower oil and in softer fats such as margarine. Many margarines are advertised as being high in polyunsaturates. This is because polyunsaturates are thought to be less potentially harmful to the heart. Doctors may recommend using margarine instead of butter or cooking with sunflower oil instead of lard to decrease the chances of a heart attack.

Top Tip
Remember that vegetable oils are high in polyunsaturates and fats are high in saturates.

Quick Test

1. What is the test for fats and oils?

2. What do fats and oils give our bodies?

3. Which provides more energy, 5 grams of sunflower oil or 5 grams of sugar?

4. How would you show that your answer to question 4 is correct?

5. Why do doctors recommend polyunsaturates in our diet?

Answers 1. Fats and oils leave greasy stains on filter paper. **2.** Energy **3.** 5 grams of sunflower oil. **4.** The oil would heat up a certain volume of water by a greater amount than the sugar would. **5.** Polyunsaturates are thought to be less damaging to the heart than saturates.

Proteins

What are proteins?

Like carbohydrates and fats and oils, proteins are also an important class of food. They are obtained from animals and plants. Proteins provide our bodies with the correct material for **body growth** and for the **repair** of body tissues. Proteins are needed if cuts and grazes are to heal properly.

You can find information about the percentage (%) of protein in different foods on page 7 of the Data Booklet.

Part of the table is shown below.

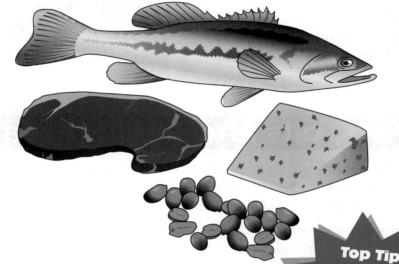

Food	Percentage (%) of protein
Peanuts	28
Cheese	25
Steak	17
Fish	16
Eggs	12
Spaghetti	10
Carrots	1
Potatoes	1
Bananas	1

Top Tip
Remember the information in the table is in your Data Booklet on page 7.

From the table you can see that foods such as nuts, cheese, meat and fish are high in proteins but fruit and vegetables are low in protein.

The parts of our bodies which have high protein content include skin, hair, nails and muscles.

Testing for proteins

When proteins are heated with a chemical known as soda lime, an alkaline gas is given off. The alkaline gas is detected by holding moist pH paper at the mouth of the test tube. The alkaline gas turns moist pH paper blue.

Carbohydrates, fats and oils contain only the chemical elements carbon, hydrogen and oxygen. Proteins contain carbon, hydrogen and oxygen too, but they also contain nitrogen. It is a compound of nitrogen which is produced when proteins are heated with soda lime, which turns moist pH paper blue.

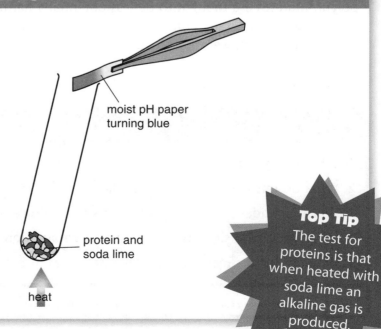

moist pH paper turning blue

protein and soda lime

heat

Top Tip
The test for proteins is that when heated with soda lime an alkaline gas is produced.

Amino acids

Proteins are polymers. This means that they are made from many monomer molecules joined together.

In the case of proteins, these monomer molecules are known as **amino acids**.

Our bodies need amino acids. We get these amino acids by eating proteins from plants and animals. When we digest our food the proteins in our food are broken down into amino acids. Our bodies then link the amino acids in a different order to make the proteins we need. In this way protein from, say, cows (in milk or cheese or steak) is converted into human protein.

The amino acids are linked in different orders to produce the different proteins in our skin and muscle. These are very different from the proteins in our hair and nails.

People who do not eat meat must make sure that their bodies get enough proteins. They can do this by eating cheese, eggs, rice and a variety of vegetables and other foods which contain proteins.

Amino acids contain the chemical elements carbon, hydrogen, oxygen and nitrogen.

One of the simplest amino acids is alanine. Its structure is

Top Tip

Remember that proteins are made up of amino acids joined together and that they contain carbon, hydrogen, oxygen **and nitrogen**.

Quick Test

1. Which chemical element is present in all proteins but not in carbohydrates or fats?
2. What is the chemical test for proteins?
3. What do our bodies use proteins for?
4. How do our bodies get the proteins they need?
5. What percentage of steak is protein?
6. Proteins are polymers. Which monomers join together to make proteins?
7. Look at the structure of alanine. Count the number of atoms of each element and write down its chemical formula.

Answers 1. Nitrogen, N. **2.** Heat the protein with soda lime. An alkaline gas will be produced. **3.** Growth and repair of tissues. **4.** From foods such as nuts, cheese, eggs, meat and fish. **5.** 17%. **6.** Amino acids. **7.** $C_3H_7O_2N$.

Fibre, vitamins and food additives

Fibre

Foods such as fruit, vegetables, bran and wholegrain bread contain fibre. The fibre in the foods we eat keeps our gut working well. In our digestive system, fibre absorbs water and swells up. This swelling provides bulk for the muscles in our gut to work on and the undigested food is squeezed along through the intestines. Without enough fibre in the food we eat, we may well get constipation.

Top Tip
Low-fibre diets can cause constipation.

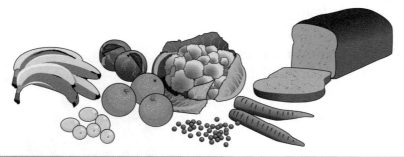

Vitamins

Vitamins are needed in very small amounts to keep our body healthy. It is possible to buy vitamins or vitamin supplements at a pharmacist's, but vitamins should be present in a healthy diet.

All vitamins are complex chemicals containing carbon and other elements but are referred to by letters, such as vitamin C.

Top Tip
Vitamins are needed to keep the body healthy.

The table below shows the types of foods which provide our bodies with vitamins A, B, C, D and E, and also the vitamin deficiency diseases which occur if our bodies do not get enough of these vitamins.

Vitamin	Foods containing this vitamin	What this vitamin does	Effects caused by lack of this vitamin
A	Cheese, eggs, oily fish, milk	Helps maintain healthy skin, immunity from infections, vision in dim light	More likely to pick up infections, night blindness
B	Meat products, eggs, cereals	Keeps nervous system healthy, helps release energy from food	Anaemia, excessive tiredness
C	Fruit and vegetables such as oranges, limes, potatoes, peppers	Helps protect cells and keeps them healthy, helps body absorb iron from food	Scurvy, bruising easily, hair loss, bleeding gums, slow-healing wounds
D	Oily fish, eggs, margarine, cereals	Keeps bones and teeth healthy	Rickets, soft bones
E	Olive oil, asparagus, nuts and seeds	Protects cell membranes	Weak muscles, tiredness

As you can see from the table, lack of vitamins in our diet can cause us to suffer poor health.

Food additives

Food additives are substances added to food in small quantities for a variety of reasons, such as making the food more nutritious or more attractive.

By law, only food additives which have been tested and approved can be added to our food. Many food additives are shown on food labels as 'E' numbers. For example, E102 is tartrazine, which is a food colouring. Some food additives, such as tartrazine, are thought to cause hyperactivity and behavioural problems in children and are no longer widely used.

Top Tip
Preservatives, colourings and flavourings are examples of food additives.

The table below gives types of food additives and reasons why they are put into foods.

Type of food additive	Reason for adding it to food
Food colourings	To make the food look more attractive and more appealing
Food flavourings	To enhance or intensify the smell and taste of the food
Food preservatives	To reduce the growth of microbes and so stop the food going bad
Vitamins and minerals	To supply these vital substances to our food in small quantities so that we do not suffer diseases caused by not having them in our diet

Utterly-Smoothier
ICE CREAM

Ingredients: Reconstituted skimmed milk, water, sugar, glucose-fructose syrup, butteroil (8·5%), whey solids, cream (2·3%), emulsifier (E471), stabilisers (locust bean gum, guar gum), flavouring, colours (annato, curcumin).

Quick Test

1. What may happen if you are on a low-fibre diet?

2. Why are vitamins needed?

3. Write down examples of foods containing vitamin A.

4. Which vitamin is found in oranges, limes and potatoes?

5. Name the disease that is caused by vitamin C deficiency.

6. Which vitamin is found in margarine?

7. Which vitamin has been missing from the diet of a child suffering from rickets?

8. Which two vitamins can you get by eating oily fish?

9. Apart from minerals and vitamins, what other food additives may be in the food that we eat?

10. What is true for all food additives before they can be put into food?

11. What may happen to jam that has not had a preservative added to it?

Answers 1. You may become constipated. **2.** To keep the body healthy. **3.** Examples include cheese, milk, eggs and oily fish. **4.** Vitamin C. **5.** Scurvy. **6.** Vitamin D. **7.** Vitamin D. **8.** Vitamins A and D. **9.** Food colourings, flavourings and preservatives. **10.** They must have been tested and approved. **11.** It may go off.

Alcohol

What are drugs?

Drugs are substances which alter the way the body works. Some drugs may damage the body, whereas other drugs, such as medicines, are helpful to the body and may at times be necessary to help recover from an illness. Some more dangerous drugs may damage our health because of how they affect our body and also our way of life.

Why is alcohol a drug?

Alcohol is a drug because it alters the way our body works. Even a small amount of alcohol is known to affect our reaction time and our concentration. This is why people should never drink alcohol if working with machinery, and why it is very dangerous to drive a car if there is alcohol in your blood.

When alcohol is taken in excess regularly it may have very harmful effects on the body. Two very important organs in the body damaged by alcohol are the liver and the brain. Taking alcohol over a long period of time can have fatal effects on these organs.

Alcohol in drinks

The alcohol in drinks is called **ethanol**. It has molecular formula C_2H_6O. This means that each molecule of ethanol contains two carbon atoms, six hydrogen atoms and one oxygen atom.

Top Tip
Remember that the alcohol in all alcoholic drinks is ethanol.

The amount of alcohol in an alcoholic drink depends on the volume and concentration of alcohol in that drink. The content of alcohol in drinks is often measured in **units**. Alcohol in drinks is broken down by the average body at a rate of about **one unit per hour**.

The approximate number of units of alcohol in drinks is shown in the table below. Remember that these are approximate and in the case of wine and spirits it will depend on the quantity measured out.

Alcoholic drink	Number of units of alcohol
One bottle of alcopop	2 units
One pint of lager	2 units
One pint of beer	2 units
One glass of wine	1 unit
One pub measure of whisky	1 unit

bottle of alcopop
2 units

pint of lager
2 units
pint of beer
2 units

glass of wine
1 unit

pub measure of whisky
1 unit

Making alcohol from fruit and vegetables

Ethanol, the alcohol present in alcoholic drinks, can be made from different fruits and vegetables in the chemical reaction known as **fermentation**.

Fruits and vegetables contain sugars and starches, and these change into alcohol during fermentation. The alcoholic drink produced depends on the source of the sugar or starch. This is shown in the table on the right.

Fruit or vegetable used	Alcoholic drink produced
Apples	Cider
Barley	Lager or beer or whisky
Grapes	Wine
Potatoes	Vodka

During fermentation, the carbohydrate glucose is changed into alcohol. To help the fermentation reaction take place, **yeast** is added. **Enzymes** present in the yeast **catalyse** the fermentation reaction. Carbon dioxide is also produced in the reaction. The carbon dioxide produced turns lime water milky, as shown in the diagram. Alcohol is poisonous, and when the concentration of alcohol is about 10–15% the yeast can no longer work properly. The concentration of alcohol in different drinks is given in the table below.

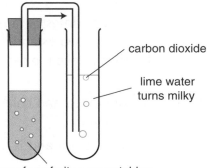

carbon dioxide

lime water turns milky

glucose from fruits or vegetables, plus yeast, changing into ethanol

Alcoholic drink	Concentration of alcohol present
Lager	4–6 %
Beer	4–6 %
Cider	5–8 %
White wine	11–14 %
Red wine	12–14 %
Whisky	35–40 %
Vodka	35–40 %

In drinks such as vodka, whisky, gin and brandy the percentage of alcohol is much higher than drinks such as beer and wine. This is because the concentration of alcohol in these drinks has been increased by a method known as **distillation**. Distillation involves boiling the fermented mixture, which is mainly alcohol and water, and evaporating off the alcohol. The alcohol vapour is then condensed back into liquid alcohol. Distillation works because ethanol and water have different boiling points. Ethanol boils at 79 °C and water boils at 100 °C. Drinks in which the alcohol and water have been partially separated by distillation are known as spirits. Whisky, gin, vodka and brandy are examples of spirits.

Top Tip
Water and ethanol can be separated by distillation because they have different boiling points.

Quick Test

1. What is the name of the alcohol in whisky, beer, vodka and other alcoholic drinks?

2. If a man has consumed 2 pints of lager, 2 measures of whisky and a bottle of alcopop, how many units of alcohol has he taken in?

3. A woman has 12 units of alcohol in her blood at midnight and then goes to her bed. How many units of alcohol will still be in her blood at 7 am?

4. What is the name of the chemical reaction in which alcohol is produced from glucose?

5. Why can distillation be used to separate alcohol from water?

Answers 1. Ethanol. **2.** 8 units. **3.** 5 units. **4.** Fermentation. **5.** Because ethanol and water have different boiling points.

Other drugs

Legal or illegal?

Medicines, alcohol, caffeine and nicotine are examples of drugs which are **legal** and can be purchased in various shops. However, since they are legal they are produced to a high degree of purity and strict rules govern their use. For example, unless you are aged 18 it is illegal for a shopkeeper to sell you alcohol. They are classed as drugs because they alter the way the body works and may have harmful effects if taken in large quantities.

Caffeine is found in coffee, tea and cola drinks. It acts as a stimulant.

Nicotine also acts as a stimulant and is found in cigarettes.

Illegal drugs include **cannabis, LSD and ecstasy**. These can have very harmful effects on the body, especially the mind.

Sometimes people feel that they cannot manage without a drug. This feeling is known as **addiction** and we say that these people are addicts. People can be addicted to legal and illegal drugs. Alcohol is a legal drug and people who are addicted to alcohol are known as alcoholics. A heroin addict is addicted to heroin, an illegal drug.

Top Tip
You should know which drugs are legal and which are illegal.

Methanol – another alcohol

The alcohol in all alcoholic drinks is ethanol. It has chemical formula C_2H_6O. In a molecule of ethanol the atoms are arranged as shown on the right.

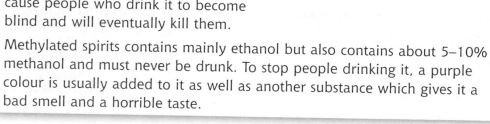

There is a simpler alcohol known as methanol. It has chemical formula CH_4O and its atoms are arranged as shown on the right.

However, methanol is very toxic. This means that it is very poisonous. It can cause people who drink it to become blind and will eventually kill them.

Methylated spirits contains mainly ethanol but also contains about 5–10% methanol and must never be drunk. To stop people drinking it, a purple colour is usually added to it as well as another substance which gives it a bad smell and a horrible taste.

Medicines

Our bodies are very complicated systems, with many different chemical reactions happening at any one time. These chemical reactions keep the body working properly.

Sometimes things go wrong and the body does not work properly. When this happens we feel ill.

We may have to take a medicine which contains the correct drugs to help the body when we are ill. The drugs in medicines are examples of legal drugs.

These drugs include mild **painkillers** such as aspirin or paracetamol, which we may take when we have a headache. Another name for painkillers is **analgesics**.

If you have an infection your doctor may prescribe an **antibiotic** such as penicillin. Antibiotics fight the micro-organisms which interfere with the chemical reactions that keep our bodies healthy.

If you have indigestion you may take an **antacid**. Antacids neutralise extra acid in our stomach. The acid helps break down carbohydrates in the food we have eaten but if our body produces too much then we can get acid indigestion or heartburn.

If you look at the label on a medicine bottle you will see that it contains many different chemicals. Some are just flavourings and sweeteners to make the medicine taste better. One of the chemicals present will work on the body to make it better. This chemical is known as the **active ingredient**.

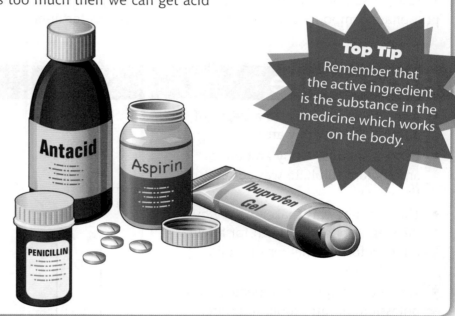

Top Tip
Remember that the active ingredient is the substance in the medicine which works on the body.

Quick Test

1. Look at the list of 6 drugs given in the box. Choose the 3 that are legal.

   > caffeine   cannabis   LSD
   > ecstasy   nicotine   alcohol

2. What does 'addiction' mean?

3. What is added to methylated spirits to discourage people from drinking it?

4. Which poisonous alcohol is present in methylated spirits?

5. What do analgesics do?

6. What do antibiotics do?

7. What does the active ingredient in a medicine do?

Answers 1. Caffeine, nicotine and alcohol are legal. **2.** Not being able to manage without a drug. **3.** A purple coloured substance and a bad tasting/smelling substance. **4.** Methanol. **5.** They are painkillers. **6.** They fight infections. **7.** The active ingredient is the substance in the medicine which works on the body to make it better.

PPA 1: Solubility

Introduction

Fertilisers added to soils help plants grow well.

To be effective a fertiliser should:

- contain one or more of the essential elements, nitrogen (N), phosphorus (P) and potassium (K).
- be soluble in water

Each of the compounds tested contained one or more of the above essential elements, but only the soluble compounds could be used as a fertiliser. In this PPA you tested some compounds to find out if they were soluble in water.

Aim

To test the solubility in water of some ammonium, potassium, nitrate and phosphate compounds to see if they could be used as fertilisers.

Procedure

- A test tube was filled to a depth of about 3–4 cm with water.
- Using a spatula, a tiny amount of ammonium sulphate was added to the water.
- The test tube was shaken gently by 'flicking' it back and forth for a few minutes to see whether the solid had dissolved.
- The above steps were repeated with potassium nitrate, sodium nitrate, ammonium phosphate and calcium phosphate.

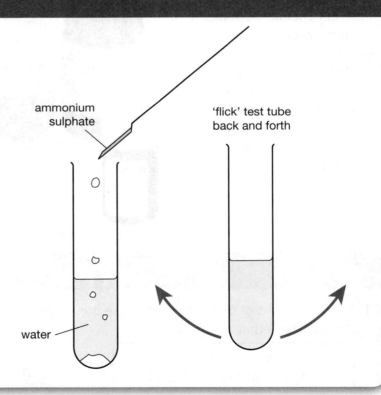

ammonium sulphate

'flick' test tube back and forth

water

Results

Name of compound	Soluble/insoluble
Ammonium sulphate	soluble
Potassium nitrate	soluble
Sodium nitrate	soluble
Ammonium phosphate	soluble
Calcium phosphate	insoluble

Conclusion

The compound which is not suitable as a fertiliser is calcium phosphate. The other compounds are all soluble and so could be used as fertilisers.

Points to note

- The compounds tested all contained one or more of the essential elements, nitrogen (N), phosphorus (P) and potassium (K).
- A very small quantity of the solid was used each time in 3–4 cm of water.
- The test tube was shaken by 'flicking' the test tube back and forth for a few minutes.
- If the solid disappeared in the water then it was soluble.
- All the compounds tested are harmful if swallowed and calcium phosphate is powdery and the dust can irritate the eyes.

Quick Test

1. What are the names and symbols of the three 'essential elements'?

2. Which two essential elements are in potassium nitrate?

3. What was the aim of the experiment?

4. Why was only a very small quantity of each solid used?

5. Which compound tested was not suitable as a fertiliser and why?

Answers 1. Nitrogen (N), phosphorus (P) and potassium (K). **2.** Potassium and nitrogen. **3.** To test the solubility in water of some ammonium, potassium, nitrate and phosphate compounds to see if they could be used as fertilisers. **4.** It would be more difficult to dissolve and if a larger quantity had been used it might have been more difficult to see if any had dissolved. **5.** Calcium phosphate, because it is insoluble in water.

PPA 2: Burning carbohydrates

Introduction

Carbohydrates are an important class of food which provide our bodies with energy. They contain the elements carbon, hydrogen and oxygen. Starch and sugar are carbohydrates.

In this experiment you burned flour as the 'starch' carbohydrate and icing sugar as the 'sugar' carbohydrate. The energy produced was used to heat water. The rise in temperature was used as an indicator of how much heat was given out.

The experiment was made fair because the same amount of flour and icing sugar were burned and the same volume of water was heated.

Aim

To show that heat energy is produced when starch and sugar are burned and to compare how much heat energy each produces when burned.

Procedure

- $10\,cm^3$ of water was measured into a boiling tube.
- The boiling tube was then clamped in a vertical position and the temperature of the water measured and recorded.
- A level spatula-full of flour was put into a Bunsen flame until the flour caught fire.
- The spatula with the burning flour was held below the boiling tube so that the flames were touching the bottom of the boiling tube.
- When the flour had stopped burning, the water inside the boiling tube was stirred and the temperature of the water measured and recorded.
- The experiment was then repeated using a level spatula-full of icing sugar in place of the flour.

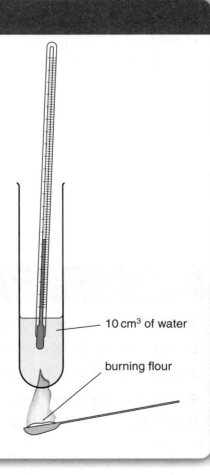

10 cm³ of water

burning flour

Results

Carbohydrate	Starting temperature of the water in °C	Final temperature of the water in °C	Rise in temperature in °C
Flour	22	30	8
Icing sugar	22	33	11

Conclusion

Both flour and icing sugar produced heat energy when burned. The icing sugar gave out more heat than the flour.

Points to note

- To make the experiment fair, things which had to be kept the same were
  (a) the volume of water heated (10 cm³)
  (b) the amount of carbohydrate burned (a level spatulaful)
  (c) the distance between the flame and the bottom of the boiling tube
- The rise in the temperature of the water gave a measure of how much heat was produced when the carbohydrate was burned.
- Care had to be taken when working with the flour, since flour dust damages the lungs if breathed in.

Quick Test

1. What was used in the experiment as the 'starch' carbohydrate?
2. What type of carbohydrate did the icing sugar represent?
3. What three things were done to make the experiment fair?
4. How was the amount of heat given out each time compared?
5. Which gives out more energy when burned, 1 gram of sugar or 1 gram of starch?

Answers 1. Flour. 2. The 'sugar' carbohydrate. 3. The volume of water, the amount of carbohydrate, the distance between the flame and the boiling tube. 4. By measuring the rise in temperature of the water. 5. 1 gram of sugar.

PPA 3: Testing for starch and sugars in food

Introduction

Starch and sugars are important carbohydrates present in foods. Our bodies use them to produce energy.

The test for starch is that it turns iodine solution from brown to a blue-black colour. If iodine solution is added to a sample of food and the iodine turns blue-black, then the food sample must contain starch.

Benedict's test can be used to find out if a food contains sugars. If a food sample is heated with blue Benedict's solution in a water bath and a cloudy orange solid is formed, then we can say that the food sample definitely contains a sugar.

If the blue Benedict's solution does not change colour, then the only sugar which may be in the food sample is sucrose. Sucrose is ordinary table sugar and it does not react with Benedict's solution. Other sugars such as maltose, fructose and glucose will react with Benedict's solution, producing a colour change from blue to cloudy orange.

Aim

To test for sugars and starch in some food samples.

Procedure

Testing for starch:

- Small samples of milk, bread, potato and egg white were added to four separate dimples on a dimple tray.
- A few drops of iodine solution were then added to each food sample and it was observed which food samples changed the iodine solution from brown to a blue-black colour.

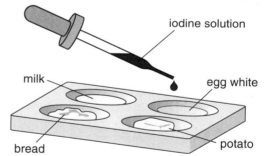

Testing for sugars:

- A large glass beaker half full of hot water was used as a hot water bath.
- Some milk was added to one test tube, followed by about 1 cm of egg white into another test tube.
- Bread crumbs were added to a third test tube and small pieces of potato added to the fourth test tube.
- Benedict's solution was then added to these four test tubes.
- The four test tubes were placed into the hot water bath and it was observed which food samples turned the Benedict's solution from blue to cloudy orange.

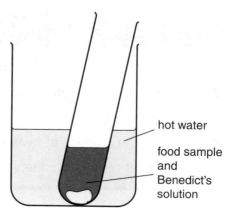

Results

Food sample	Observations on adding iodine solution	Observations on heating with Benedict's solution
Milk	No change, iodine solution stays brown	Blue Benedict's solution turns cloudy orange
Bread	Iodine solution changes from brown to blue-black	Blue Benedict's solution turns cloudy orange
Potato	Iodine solution changes from brown to blue-black	No change, Benedict's solution stays blue
Egg white	No change, iodine solution stays brown	No change, Benedict's solution stays blue

Conclusion

The completed table for the conclusion is:

Food sample	Is starch present?	Are sugars present?
Milk	No	Yes
Bread	Yes	Yes
Potato	Yes	No
Egg white	No	No

Points to note

- Foods containing starch will turn iodine solution from brown to blue-black.
- Benedict's solution will change from blue to cloudy orange when warmed in a water bath with food containing sugars.
- Bread contains both starch and sugars
- Care has to be taken when using Benedict's solution and iodine solution because both are harmful substances.

Quick Test

1. How would you test a food sample to see if it contained starch?
2. How would you test a food sample to show that it contains sugars?
3. Which food sample contained starch but no sugars?
4. Which food sample contained neither starch nor sugars?
5. Which food sample contained sugars but no starch?

Answers 1. Add iodine solution to the food sample. If the iodine solution turns from brown to blue-black then the food contains starch. **2.** Warm the food sample with Benedict's solution in a hot water bath. The Benedict's solution will turn from blue to cloudy orange showing that sugars are in the food. **3.** Potato. **4.** Egg white. **5.** Milk.

Glossary

Acid   A substance that forms a solution with a pH of less than 7.

Acid rain   Rain containing dissolved sulphur dioxide and/or nitrogen dioxide. It has damaging effects on buildings, structures made of iron or steel, soils and plant and animal life.

Active ingredient   The chemical in a medicine which works on the body to make it better.

Addiction   Being unable to manage without a drug.

Air   A mixture of gases - approximately 80% nitrogen and 20% oxygen.

Alkali   A substance that forms a solution with a pH of more than 7.

Alloy   A mixture of metals or a mixture of metals and non-metals. Brass, solder and 'stainless' steel are examples of alloys.

Amino acids   The small monomer units used in making proteins.

Antibiotics   Drugs that fight micro-organisms which interfere with the chemical reactions that keep our bodies healthy. Many antibiotics contain penicillin.

Atomic number   A number given to each element in the Periodic Table.

Atoms   The tiny particles that make up elements.

Battery   A collection of two or more cells joined together.

Benzene   A toxic hydrocarbon that used to be present in unleaded petrol.

Biodegradable   Able to be broken down by bacteria in the soil and eventually rot away.

Biogas   A renewable fuel formed by the breakdown of plant and animal waste. It consists mainly of methane.

Burning   A chemical reaction in which a substance reacts with oxygen in the air and produces heat energy.

Carbohydrates   Compounds which contain carbon, hydrogen and oxygen and are used by the body to produce energy.

Carbon monoxide   A poisonous gas formed when fuels burn in a limited supply of air.

Catalyst   A substance which speeds up a chemical reaction and is not used up by the reaction.

Catalytic converters   A device which is fitted to exhausts to speed up the conversion of pollutant gases into less harmful gases.

Cell   A device which produces electricity from a chemical reaction.

Chemical formula   A shorthand way of representing a substance. It shows what elements are present in the substance and the number of atoms of each e.g. CO_2 is the formula for carbon dioxide and shows that one molecule of carbon dioxide contains one carbon atom and two oxygen atoms.

Chemical reaction   A process in which substances change to form one or more new substances.

Chemical symbol   A shorthand way of representing an element. It consists of one or two letters e.g. C for carbon and Co for cobalt.

Chlorophyll   A green-coloured compound present in the green parts of plants. It absorbs light energy needed for photosynthesis.

Cleaning chemicals   Substances like soaps, detergents, washing-up liquids, washing powders and shampoos. They clean surfaces because they are soluble in oil and grease as well as in water.

Coal   A solid fossil fuel which is mainly carbon.

Combustion   Another word for burning.

Complete combustion   When a substance burns completely in a plentiful supply of air.

Compound   A substance in which two or more elements are chemically joined.

Concentrated solution   A solution in which a large amount of substance (solute) has been dissolved.

Corrosion   A chemical reaction that takes place on the surface of a metal. The metal reacts with substances in the air to form a compound. Iron rusting is an example of corrosion.

Cracking   A chemical reaction in which large hydrocarbons are broken down into a mixture of smaller, more useful hydrocarbons.

Crude oil   A liquid fossil fuel which contains a mixture of hydrocarbons.

Diatomic molecule   A molecule which contains only two atoms.

Digestion   The breaking down of large food molecules into smaller ones so that they can pass into the blood stream. For example, starch is digested into glucose and proteins are digested into amino acids.

Dilute solution   A solution in which a small amount of substance (solute) has been dissolved.

Distillation   A method of separating liquids which have different boiling points. It is used to increase the alcohol concentration of fermented mixtures.

Drugs   Substances which alter the way the body works.

Dry-cleaning   A process of cleaning clothes using special solvents that are good at dissolving stains.

Dyes   Coloured compounds which are used to give bright colours to clothing.

Electroplating iron   A process in which a thin layer of a metal is coated on to iron using electrolysis. The metal coating prevents rusting.

Element   A substance which cannot be broken down into a simpler substance. All the elements are listed in the Periodic Table on page 8 of your Data Booklet.

Enzyme   A catalyst which speeds up chemical reactions in living things.

Essential elements   These are nitrogen (N), phosphorus (P) and potassium (K) and are needed for healthy plant growth.

Ethanol   The alcohol present in alcoholic drinks. It is a drug and if taken in excess can have harmful effects on the body. It can be made by the fermentation of glucose.

Fats and oils   Compounds which contain carbon, hydrogen and oxygen and are used by the body to produce energy. They are more concentrated sources of energy than carbohydrates.

Fermentation   A chemical reaction in which glucose is broken down into ethanol and carbon dioxide. It is catalysed by enzymes present in yeast.

Fertiliser   A natural or artificial substance which is added to the soil to restore essential elements.

Fibre   An important part of our diet because it keeps the gut working well and prevents constipation.

Fibres   Thin thread-like strands used to make clothing fabrics. They are made up of long chain molecules called polymers.

Finite resources   Ones that will run out and can never be renewed. Examples include the fossil fuels.

Fire triangle   An aid to understanding how fires can be put out. Take away the fuel or the oxygen or the heat from a fire and it goes out.

Flammability   A measure of how easily a substance catches fire.

Food additives   Chemicals like vitamins, minerals, colourings, flavourings and preservatives which are added to food. They must first be tested and approved.

Fossil fuels   They are coal, oil, natural gas and peat and were formed over millions of years from the remains of dead animals and plants.

Fraction   A mixture of hydrocarbons with boiling points within a certain range of temperature.

Fractional distillation   The process used to separate crude oil into fractions according to the different boiling points of the hydrocarbons in the fractions.

Fuel   A substance which burns to produce heat energy.

Fuel crisis   A problem caused by not having enough fuel supplies to meet demand.

Fungicides   Chemicals which prevent diseases in plants caused by bacteria and fungi.

Galvanising   A process in which iron is coated with zinc in order to prevent rusting.

Global warming   Increasing amounts of carbon dioxide in the atmosphere has caused more of the Sun's energy to be absorbed and has increased the temperature of the atmosphere.

Greenhouse effect   This is caused by carbon dioxide in the atmosphere which absorbs some of the Sun's energy and keeps the Earth warm.

Hazard symbols   Labels put on containers to indicate they contain dangerous chemicals. Hazard symbols include those for harmful/irritant, toxic, corrosive and flammable substances.

Herbicides   Chemicals which kill weeds.

Hydrocarbon   A compound which contains hydrogen and carbon only.

Illegal drugs   These include cannabis, LSD and ecstasy.

Incineration   Disposing of waste, such as used plastics, by burning.

Incomplete combustion   When a substance burns in a limited supply of air.

Insoluble   When a substance (solute) does not dissolve in a liquid (solvent), it is said to be insoluble.

Ionic compounds   Compounds made up of oppositely charged ions. They have high melting and boiling points because the bonds between ions are very strong and they conduct electricity when dissolved in water and when molten.

Ions   Tiny particles like atoms but they have a positive or negative charge.

Kevlar   A plastic which is used in bullet-proof vests because it is very strong and in fire-fighters' clothing since it is heat resistant.

Legal drugs   These include medicines, alcohol, nicotine and caffeine.

Malleable   Can be beaten into different shapes. Metals are malleable.

Medicines   They contain drugs which help the body when it is not working properly.

Metal   A type of element. All metals conduct electricity. They lie below and to the left of the heavy black line on the Periodic Table on page 8 of your Data Booklet.

Metal reactivity   A measure of how fast a metal reacts with other substances. A metal reactivity series can be found on page 6 of your Data Booklet.

Methanol   An alcohol which is very toxic. It can cause blindness and death and is present in methylated spirits.

Methylated spirits A liquid mixture which contains mainly ethanol but also some methanol. It has a colour and bad tasting substance added to it to discourage people drinking it.

Minerals Compounds in the diet which supply us with important elements, e.g. calcium for bones and teeth and iron for blood.

Mixture Two or more substances just mixed together but not chemically joined.

Molecular substances Substances which are made up of molecules. They have low melting and boiling points because the bonds between molecules are weak and they do not conduct electricity.

Molecule A group of two or more atoms which are held together by strong bonds.

Natural fibres Fibres that come from plants and animals. Examples include silk, wool and cotton.

Neutral solutions A solution with a pH = 7.

Neutralisation A chemical reaction which takes place when acids react with alkalis or metal carbonates. Water and a salt are always formed in neutralisation reactions.

Nitrogen dioxide A poisonous gas formed when nitrogen and oxygen in the air react during lightning storms and around the spark in petrol engines. It causes acid rain.

Non-metal A type of element. All non-metals except carbon (graphite) do not conduct electricity. They lie above and to the right of the heavy black line on the Periodic Table on page 8 of your Data Booklet.

Periodic Table An arrangement of the elements in order of increasing atomic number. All the elements in each column of the Periodic Table have similar chemical properties.

Pesticides Chemicals used to kill and control pests like insects, slugs, snails etc.

pH A number that indicates how acidic or alkaline a solution is.

pH scale A scale which runs from below 0 to above 14 and is a measure of how acidic or alkaline a solution is. Acids have a pH below 7 and alkalis have a pH above 7.

Photosynthesis A process which takes place in the leaves of plants and involves the conversion of carbon dioxide and water into glucose and oxygen. Light energy is needed for photosynthesis to take place.

Plastics Synthetic materials which are polymers.

Pollutant Any substance which damages the environment and harms living organisms within the environment.

Polymer A very large molecule which is formed when hundreds of small monomer molecules join together.

Polymerisation A chemical reaction in which small monomer molecules are converted into a large polymer molecule.

Polyunsaturates Compounds found mainly in oils. They are thought to be less harmful to the heart than saturates.

Proteins Compounds which contain carbon, hydrogen, oxygen and nitrogen and used for the growth and repair of body tissues.

Reaction speed How fast a chemical reaction takes place. It can be affected by changes in particle size, temperature, concentration and the use of a catalyst.

Recycling Reprocessing plastic waste into useful products.

Renewable resources Resources that will not run out and can be remade. Examples include methane, ethanol and hydrogen.

Respiration A process whereby all living things get energy from glucose. It involves the reaction between glucose and oxygen to form carbon dioxide and water.

Root nodules Lumps on the roots of plants, such as peas, beans and clover, containing bacteria which convert nitrogen from the air into nitrates.

Rust indicator A solution which turns blue if rusting occurs.

Rusting The corrosion of iron. Both oxygen and water must be present for iron to rust.

Salt A compound formed in a neutralisation reaction e.g. sodium chloride, potassium sulphate, copper nitrate etc.

Saturated solution A solution in which no more substance (solute) will dissolve.

Saturates Compounds found mainly in fats. They are believed to increase the cholesterol level in the blood which is thought to be responsible for heart disease.

Scum A grey solid which is formed when soap is added to hard water.

Signs of chemical reaction These include a colour change, a precipitate forming, a gas being given off and a temperature change.

Silicones Plastics which repel water and are used in water-proofing clothes and as sealants in baths and showers.

Soluble When a substance (solute) dissolves in a liquid (solvent), it is said to be soluble.

Solution A mixture formed when a substance (solute) dissolves in a liquid (solvent).

Soot Particles of carbon which are formed when petrol and diesel undergo incomplete combustion.

Starch A polymer carbohydrate made in plants from glucose monomer units. It does not readily dissolve in water and is not sweet. It is used as an energy store in plants.

Sugars Small carbohydrate molecules which are soluble in water and sweet in taste. Examples include glucose, fructose, maltose and sucrose.

Sulphur dioxide A poisonous gas formed when fossil fuels are burned and causes acid rain.

Synthetic fibres Fibres that are manufactured by the chemical industry. Examples include nylon (a polyamide) and Terylene (a polyester).

Thermoplastic plastic A plastic that softens on heating and can be reshaped.

Thermosetting plastic A plastic which does not soften on heating and cannot be reshaped.

Tin-plating iron A process in which iron is coated with tin in order to prevent corrosion.

Toxic Poisonous

Trace elements Elements which are needed by the body only in tiny amounts. They are supplied by minerals in the diet.

Unit of alcohol A measure of the alcohol content of alcoholic drinks. Approximately one unit of alcohol is contained in half a pint of beer, a pub measure of a spirit and a glass of wine. Alcohol is broken down in the body by about one unit per hour.

Viscosity A measure of the thickness or runniness of a liquid.

Vitamins Complex chemicals required to keep the body healthy.

Word equation A way of describing a chemical reaction showing the names of the reactants and products e.g. sulphur + oxygen → sulphur dioxide.

Chemical tests

Acids   turn pH paper or universal indicator solution red.

Alkalis   turn pH paper or universal indicator solution blue/purple.

Carbon dioxide   turns limewater milky.

Fats and oils   leave a greasy see-through stain on filter paper.

Glucose or fructose or maltose   turn warm blue Benedict's solution cloudy orange.

Hydrogen   burns with a squeaky pop.

Oxygen   relights a glowing splint.

Proteins   produce a gas which turns moist pH paper blue when heated with soda lime.

Rust   turns rust indicator blue.

Starch   turns brown iodine solution blue/black.

Intermediate 1 Chemistry
Index